Mechanical Engineering Series

Series Editor

The Mechanical Engineering Series presents advanced level treatment of topics on the cutting edge of mechanical engineering. Designed for use by students, researchers and practicing engineers, the series presents modern developments in mechanical engineering and its innovative applications in applied mechanics, bioengineering, dynamic systems and control, energy, energy conversion and energy systems, fluid mechanics and fluid machinery, heat and mass transfer, manufacturing science and technology, mechanical design, mechanics of materials, micro- and nano-science technology, thermal physics, tribology, and vibration and acoustics. The series features graduate-level texts, professional books, and research monographs in key engineering science concentrations.

More information about this series at http://www.springer.com/series/1161

Esther Titilayo Akinlabi ·
Rasheedat Modupe Mahamood

Solid-State Welding: Friction and Friction Stir Welding Processes

Springer

Esther Titilayo Akinlabi
Department of Mechanical
Engineering Science
University of Johannesburg
Johannesburg, Gauteng, South Africa

Rasheedat Modupe Mahamood
Department of Mechanical
Engineering Science
University of Johannesburg
Johannesburg, Gauteng, South Africa

Department of Mechanical Engineering
University of Ilorin
Ilorin, Nigeria

ISSN 0941-5122 ISSN 2192-063X (electronic)
Mechanical Engineering Series
ISBN 978-3-030-37014-5 ISBN 978-3-030-37015-2 (eBook)
https://doi.org/10.1007/978-3-030-37015-2

This Springer imprint is published by the registered company Springer Nature Switzerland AG
The registered company address is: Gewerbestrasse 11, 6330 Cham, Switzerland

This book is dedicated to Almighty God.

Foreword

Solid-state welding processes are important welding process that has helped to offset some of the problems caused by the fusion welding processes such as porosity and difficulty joining dissimilar materials. The porosity is caused as a result of melting of the welded surfaces which can be a result of the gas bubble in the weld pool during the welding process. The solid-state welding does not involve the melting of the materials hence the gas porosity is avoided. Also, because the solid-state welding processes do not need the use of filler material which will be adding to the overall weight of the components being joined, makes friction and friction stir welding process a choice welding technique where weight saving is important such as in the automobile and aerospace industries due to their ability to reduce the buy-to-fly ratio in the aerospace industry. Friction stir welding processing is also an important solid-state material modification process that is used in surface modification as well as for bulk material. This book titled *Solid-State Welding: Friction and Friction Stir Welding Processes* was writing by the authors to highlight the importance of these solid-state manufacturing processes. The Fourth Industrial Revolution is upon us and the manufacturing process that is not compatible or Industry 4.0 ready will lose its relevance in this digital world. The book also highlights the importance of friction stir welding in this industrial revolution and what is needed to be done in terms of research to better position the manufacturing process. I am also an expert in this field and has published a number of articles in this field, I am in a good position to recommend this book to you. This book will benefit the upcoming researchers, practitioners and industries by helping them to understand the capability of the friction welding, friction stir welding process and friction stir processing. This book will be of great benefit to readers.

The authors of this book have made a great contribution to this field of research with lots of published works that include journals articles, conference proceedings, book chapters and book. The authors are qualified to write this book as experts in this field. There are seven chapters in this book. The book started with the introduction of the friction welding, friction stir welding and friction stir processing. This chapter provided adequate background knowledge of the friction welding, friction stir welding and friction stir processing for the readers as well as the new researchers in this field for an adequate understanding of the these solid-state welding and processing. The principle of operation, the advantages, limitations and areas of application of these processes are introduced and highlighted in this chapter. Friction welding process is the focus of Chap. 2. This solid-state welding process is described in detail. The working principle of the friction welding process is explained in this chapter. The processing parameters that affect the friction stir welding process are explained in this chapter. Advantages, limitations, areas of application and some of the research work in this field are also presented. The friction stir welding process is presented in Chap. 3. The principle of operation is fully described. The processing parameters that govern the welding process are explained. The advantages, the limitations and areas of application of friction stir welding process are also presented. Some of the research progress in the friction stir welding process is also presented in this chapter. The friction stir processing, a material property modification technique, is described in Chap. 4. The processing parameters that govern this property modification process are also highlighted in this chapter. Advantages, disadvantages and areas of application of friction stir processing are also presented in this chapter. Case studies of friction stir welding and friction stir processing are described in Chaps. 5 and 6 in this book. The future research direction in friction welding, friction stir welding and friction stir processing are explained in Chap. 7. The Fourth Industrial Revolution and cloud computing in friction stir welding process are also presented in this chapter and the chapter ends with a summary of the book.

Haven gone through this book and being an expert in this field of research, I hereby strongly recommend this book to the readers because of the great benefits the readers can gain from this book. This book will also help the new researchers in this field to understand the basic principles of these solid-state material processing and the state of the art of the processes in order to contribute to the process of these great technologies.

Prof. Satish Vasu Kailas
Surface Interaction and Manufacturing Laboratory
Department of Mechanical Engineering
Indian Institute of Science
Bangalore, India

Preface

The friction welding, friction stir welding and the friction stir processing are solid-state manufacturing processes that have great potentials for the manufacturing industries. The joining of dissimilar materials using the traditional fusion welding technique has always been challenging usually as a result of wide apart the melting temperature of the materials which create segregation problems when the materials solidify and hence degrading the integrity of the weld produced. Surface modification has become a growing need in the manufacturing world especially when the materials need to come in rubbing contact with another material and also when an improved corrosion resistance property is required. This book began with the introduction and background of the friction welding, friction stir welding and friction stir processing. A detailed introduction of each of these solid-state processes was presented in this first chapter of the book. The basics principle of each of the processes is explained with advantages, limitations and areas of application in Chap. 1. The friction welding is presented in detail in Chap. 2. The working principle of friction welding and the processing parameters that influence the properties of friction welding is fully described in this chapter. Advantages, limitations and areas of application of friction welding are also described. The research progress in friction welding was also presented. The friction stir welding process is described in detail in Chap. 3. The friction stir processing is described in Chap. 4 while the case studies in friction stir welding process and friction sir processing are presented in Chaps. 5 and 6. The future research needs in friction welding, friction stir welding and friction stir processing are presented in Chap. 7. This book is organized as follows:

Chapter 1—Introduction and brief background of the friction welding, friction stir welding and friction stir processing are discussed in this chapter.

Chapter 2—Friction welding is an important solid-state welding process that can be used to join solid and hollow pipes as well as irregular shapes that have helped to join different materials and very useful in automobile and aerospace industry. The working principle friction welding, the processing parameters that influence the process, advantages, limitations and areas of application of friction welding are explained in this chapter. The research progress in this field is also highlighted in this chapter.

Chapter 3—The full description of the working principle of the friction stir welding process is presented in this chapter. The processing parameter that governs the process, the advantages, disadvantages and areas of application are also presented in this chapter. The current research in friction stir welding is also highlighted in this chapter.

Chapter 4—The description of the principle of operation of friction stir processing is presented in this chapter. The processing parameters that have great influence on the resulting properties in friction stir processed material are also presented in this chapter. Advantages, limitations and areas of application of friction stir processing are also highlighted in this chapter. Some of the research work in friction stir processing are also presented.

Chapter 5—Case study of friction stir welding of copper and aluminium is presented in this chapter.

Chapter 6—Case study of friction stir processing of pure aluminium is presented in this chapter.

Chapter 7—This chapter presents the future research direction in friction welding, friction stir welding process and friction stir processing. The Fourth Industrial Revolution and the cloud computing in friction stir welding process is also presented. The summary of the whole book is also presented in this chapter.

Johannesburg, South Africa

Esther Titilayo Akinlabi
Rasheedat Modupe Mahamood

Acknowledgements This work was supported by the University of Johannesburg and University of Ilorin.

Book Description

This book titled *Solid-State Welding: Friction and Friction Stir Welding Processes* presents the critical information on the working principles of solid-state material processing techniques—friction welding, friction stir welding and friction stir processing with illustrations. The application of these solid-state processes is explained with the advantages they offer and limitations of the processes. One of the attractive features of these technologies is the ability to join dissimilar materials with great success are highlighted in this book. The processing parameters that govern these solid-state processes are explained. The importance of the friction stir welding process in the Fourth Industrial Revolution and the importance of cloud computing in the friction stir welding process are also presented. The use of friction stir processing in surface engineering for the surface modification of materials is also presented in this book. Case studies in friction stir welding and friction stir processing are also presented for joining dissimilar materials of aluminium and copper and in the processing of pure aluminium. The research progress in each of these filed also presented in this book with an extensive bibliography in this research field are contained in this book. This book explains the future research needs in each of these solid-state material processing techniques.

Contents

Chapter 1
Introduction to Friction Welding, Friction Stir Welding and Friction Stir Processing

1.1 Introduction

Joining of materials are constantly needed at one point or another during the developmental stages of materials as well as for repair of damaged parts. The two main types of joining are permanent and temporary joining processes. The temporary joining process includes fasteners, cotter joint, knuckle joints, etc. Permanent joining techniques include revet joint, adhesive joint and welding. Welding is a permanent joining process that joins two or more solid components together to form a unit component that cannot be separated without damaging the components. Welding offers many advantages over the other permanent joining processes. Welding can be defined as one of the manufacturing processes, where two or more similar or dissimilar materials are joined together permanently forming coalescence with or without the applications of heat, filler material or external pressure. There are two main classes of welding processes, namely fusion welding and solid-state welding. Fusion welding is the process of melting down the faying surfaces of parent materials with the filler material to form weld bead. Examples of fusion welding include arc welding, gas welding and intense energy beam welding. Solid-state welding is achieved without the melting of the materials to be joined and the welding takes place in the solid state of the materials. In solid-state welding, application of heat is not required but pressure application may be necessary to produce a sound joint. Sometime, the base materials may be heated to an elevated temperature during the joining process but the temperature is always lower than the melting points of these materials. Explosion welding, ultrasonic welding, friction welding and friction stir welding processes are some of the examples of solid-state welding processes. The focus of this book is on friction welding, friction stir welding and a surface modification process that is based on friction stir welding process, friction stir processing are described in this book. Each of these welding processes is introduced in this chapter and they are presented in the following subsections.

E. T. Akinlabi and R. M. Mahamood, *Solid-State Welding: Friction and Friction Stir Welding Processes*, Mechanical Engineering Series,
https://doi.org/10.1007/978-3-030-37015-2_1

1.2 Introduction to Friction Welding

Technological development has evolved over time with the need to produce devices using different materials and joining of dissimilar materials has become inevitable. Joining of two dissimilar materials using the traditional welding technique has always posed problems such as porosity and other forms of welding defect due to the mismatch in the materials'properties. This is why the need for alternative welding process such as friction welding was developed. Friction welding process utilizes the heat generated from the relative movement of the materials to be joined in order to plasticize the material around the contact zone and in combination of adequate pressure to produce a metallurgically bonded material.

Friction welding was first tried in the fifteenth century with the first patent granted to a machinist, J.H. Bevington. Bevington applied friction welding in joining of metal pipes [1]. Friction welding was initially applied to cutting tools in the metal processing industry and the areas of application have grown since the inception to date. Friction welding was applied to plastic materials in the 1940s in the USA and Germany [1, 2]. The first studies on friction welding in England were carried out by the Welding Institute in 1961. The Caterpillar Tractor Company in the USA modified the friction welding process and developed the method of inertia welding in 1962. Friction welding process has since found several applications throughout the world.

Friction welding is a solid-state welding process that uses thermal energy that is produced by friction between the interface of parts to be welded as a result of relative motion.

The pressure is applied simultaneously in addition to the relative motion of the part, which causes the heat to be generated as a result of friction. The generated heat is used to raise the temperature of the surfaces to be welded until the plastic forming temperature is achieved. The relative motion between the parts to be welded is achieved by holding one of the two parts in a stationary position while the other one is rotated against the stationary one with axial pressure applied. After a sufficient length of time is reached, and the rotational speed reaches the desired value, then, the rotation is stopped. Further pressure is applied and for a sufficient length of time to allow the pieces to be metallurgically bounded together.

The two main types of friction welding based on the type of motion involved in the welding process are rotary motion and linear friction motion. The rotary friction welding is best suited for joining circular parts such as rod, circular bar, tube and pipe while the linear friction welding is suited for other geometry. The rotary friction welding is based on rotary or circular motion while the linear friction welding is based on linear reciprocating motion. This welding process can be used to joint bars, rod and tubes of metal of more than 100 mm in diameter. Both types of friction welding processes are based on the same working principle with the only difference being the type of motion that is used to achieve the welding process. Reciprocating motion is used in the linear friction welding while rotary motion is used in the rotary friction welding. The working principle of the rotary friction welding process is described as follows:

- Before the commencement of the friction welding operation, the two pieces of metal to be joined are prepared in order to have a smooth square surface. One of two workpieces is mounted on a rotor driven chuck while the other workpiece is held on a fixture. The rotor is allowed to rotate at a high speed against the stationary workpiece. A little pressure is applied on the stationary workpiece that allows the cleaning of the two surfaces through burnishing action.
- After the initial surface preparation is completed, a high-pressure force is applied to the stationary workpiece, which forces it toward the rotating workpiece that then generates a high friction force. This high friction, in turn, generates heat at the interface of the two workpieces. This process continues until the plastic forming temperature is achieved.
- The rotor is then stopped and the applied pressure force is increased gradually until the whole weld is formed. The stages involved in friction welding are shown in Fig. 1.1.

The friction welding can also be classified as a continuous friction welding or an inertial friction welding. The continuous induce friction welding is the same as the one described above with the rotor connected with a band brake. The band brake is used to stop the motion when the plastic temperature limit is reached. The inertial friction welding on the other hand uses flywheel instead of the band brake. The flywheels are connected to the motor with the shaft at the beginning, and when the speed or friction limit is reached, the engine flywheel is separated from the shaft flywheel which causes the motion of the shaft flywheel to stop without applying any brake. Since the inception of this important solid-state welding process, a number of research works have been reported in the academic literature to gain more understanding of the process and also to improve the performance of the process [3–10].

Friction welding found its application in many industries that include: automobile, marine, oil and chemical industries. A number of components has been successfully joined using the friction welding process, such as, for welding tubes and shafts, gears, valves, axle tube, shaft, hydraulic piston rod, truck rollers bushes, for welding copper and aluminium equipment in electrical and electronic industries and connecting rod. Some of the components that are joined using friction welding process is shown in Fig. 1.2.

Advantages of friction welding include narrow heat-affected zone, high strength weld, it is a green energy process and environment-friendly process because it requires less energy and the process does not generate smoke which is harmful to man and the environment. The process can be easily automated, it produces high efficient welding, and variety of similar and dissimilar metals can be welded together using the friction welding process.

Some of the limitations of this welding process include: it is mostly used for round bars of same cross section and simple shapes, non-forgeable material cannot be weld, it requires high setup capital, preparation of workpiece is more critical and it can only be used for smaller parts of machines, big parts cannot be friction welded because of the limitation in machine that can support such big part.

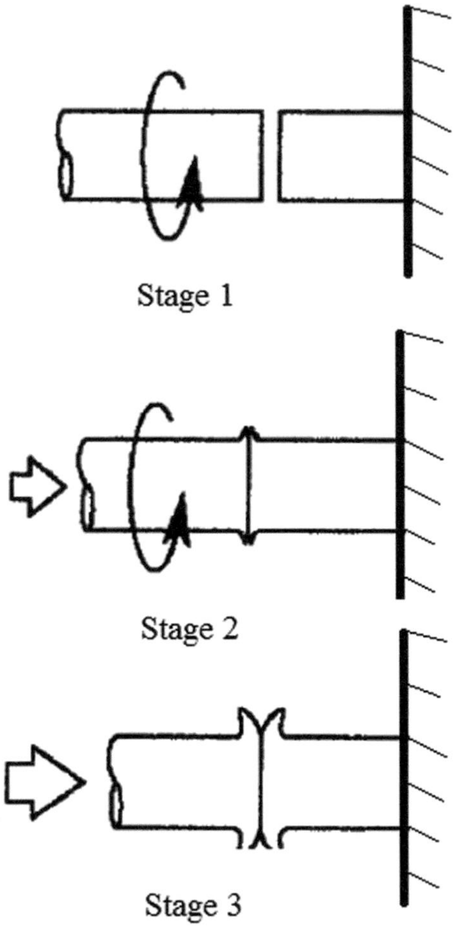

Fig. 1.1 The three stages in friction welding process

1.3 Introduction to Friction Stir Welding Process

Friction stir welding (FSW) is another solid-state welding and it is similar to the friction welding process because it is also based on the principle of friction. Friction stir welding process is a relatively new welding process that has successfully been commercialized within the short space of time from its inception. Friction stir welding was invented in 1991 by The Welding Institute (TWI) [11]. TWI has since then, continued in various research projects to further develop the technology for welding aluminium and other difficult to weld metals [11–27]. Since when FSW has been invented, the welding process has received worldwide acceptance and great attention in the research community. Friction stir welding is an effective technique for welding aluminium, copper, brass, as well as other low-melting alloys. A number of research works have been conducted to develop tools that can withstand high temperature and

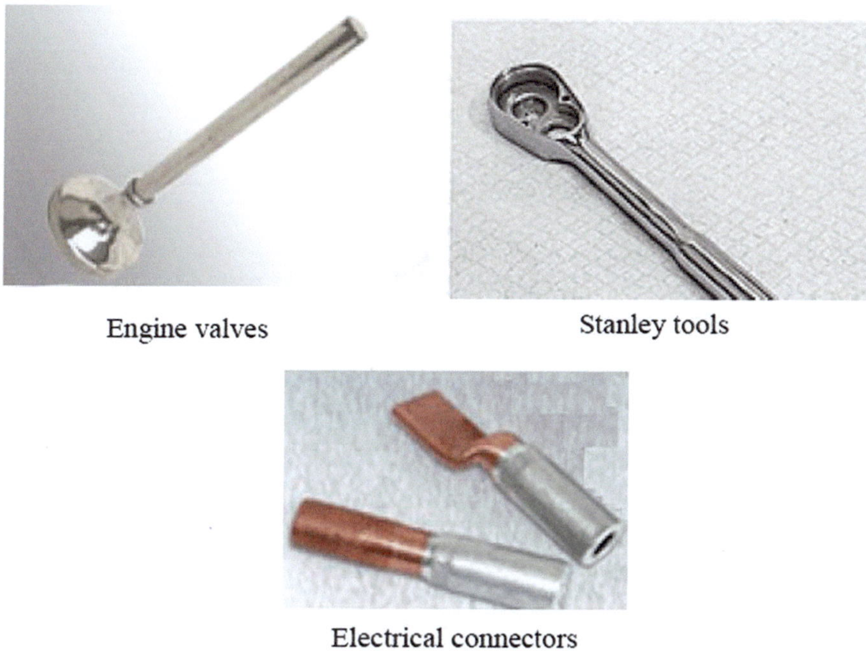

Fig. 1.2 Pictures of friction-welded components

pressure for FSW of material with a high melting point, such as, stainless steel and nickel-based alloys. The developed new tool materials include the polycrystalline cubic boron nitride (PCBN), ceramic and tungsten rhenium. Also, the development of liquid-cooled tool holder and telemetry system has been a major advancement to further expanding the process capability as well as expanding the application area of the friction stir welding process. The rapidly increasing knowledge in the material developments and the availability of the enabling technologies that made the development of special tool materials possible has also been the key driver in advancing this amazing welding process.

The principle of operation of friction stir welding process involves the use of a cylindrical non-consumable tool with shouldered and a profiled pin, that is, rotated and plunged slowly into the parts to be joined. After the tool is plunged into the two pieces of sheet or plate material to be joined together, the tool is continuously rotated under pressure and also travel through the workpiece. By so doing, the tool rotation and travel cause friction that generates heat in the material. The heat generated softens the materials and the two materials are stirred and mixed together as the welding process progresses. The mixing of the materials takes place in the solid state because the materials are only softened with the heat generated but the heat is not sufficiently large to melt the material. This action produces a sound metallurgically bonding between the two materials being joined together. The parts to be joined must be clamped onto a backing bar so as to prevent the faces to be joined from being forced

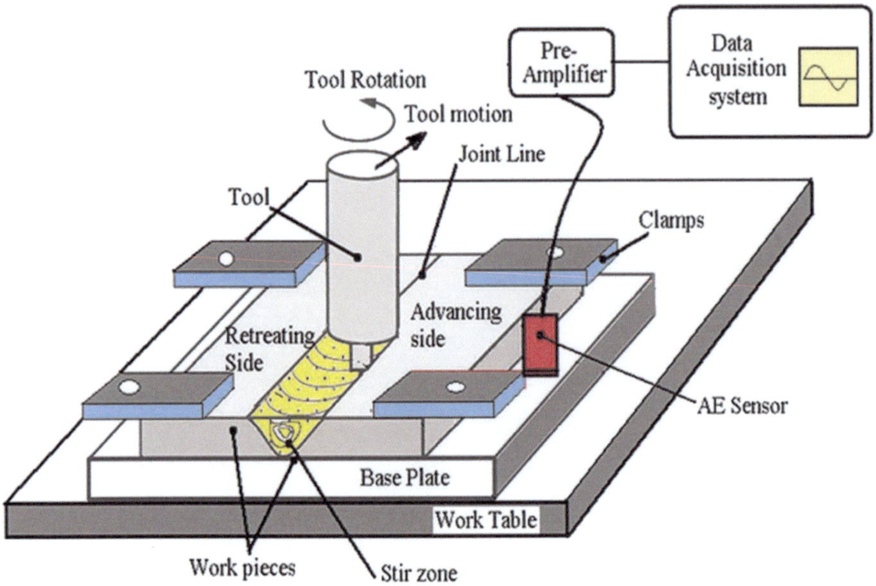

Fig. 1.3 Schematic diagram of the experimental setup for friction stir welding [12]

apart or shifted out of position during the welding process as a result of high rotation and high pressure that is involved in this welding process. The schematic diagram of friction stir welding process is shown in Fig. 1.3 [12]. Some of the different tool profiles that can be used in friction stir welding process are shown in Fig. 1.4.

The plasticized material between the interface of the materials being welded is transferred from the leading edge of the tool to the trailing edge of the tool probe and the materials are forged together through the contacting tool shoulder and the tool pin profile (see Fig. 1.4) leaving a solid-phase bond between the two materials being joined.

Friction stir welding is an excellent manufacturing process with lots of advantages, some of these advantages include it produces excellent weld quality, with low distortion, low shrinkage, no porosity because there is no melting of the base materials, no lack of fusion, no change in material composition because there is no filler material used in the welding process. Other advantages of the FSW include it is an economical welding method because it uses non-consumable tool, it requires minimal surface preparation, usually degreasing, it uses low energy input when compared to the traditional welding methods such as arc welding process, it produces no fumes or toxic gases are produced which can be harmful to the operator and the environment, the weld has a good surface finish with no welding spatter produced and dissimilar materials and alloys can be joined together such as copper and aluminium. Limitations of the FSW process include low mechanical stability of the tool at the operating temperature and it cannot be used for the traditional tee filet joint configuration.

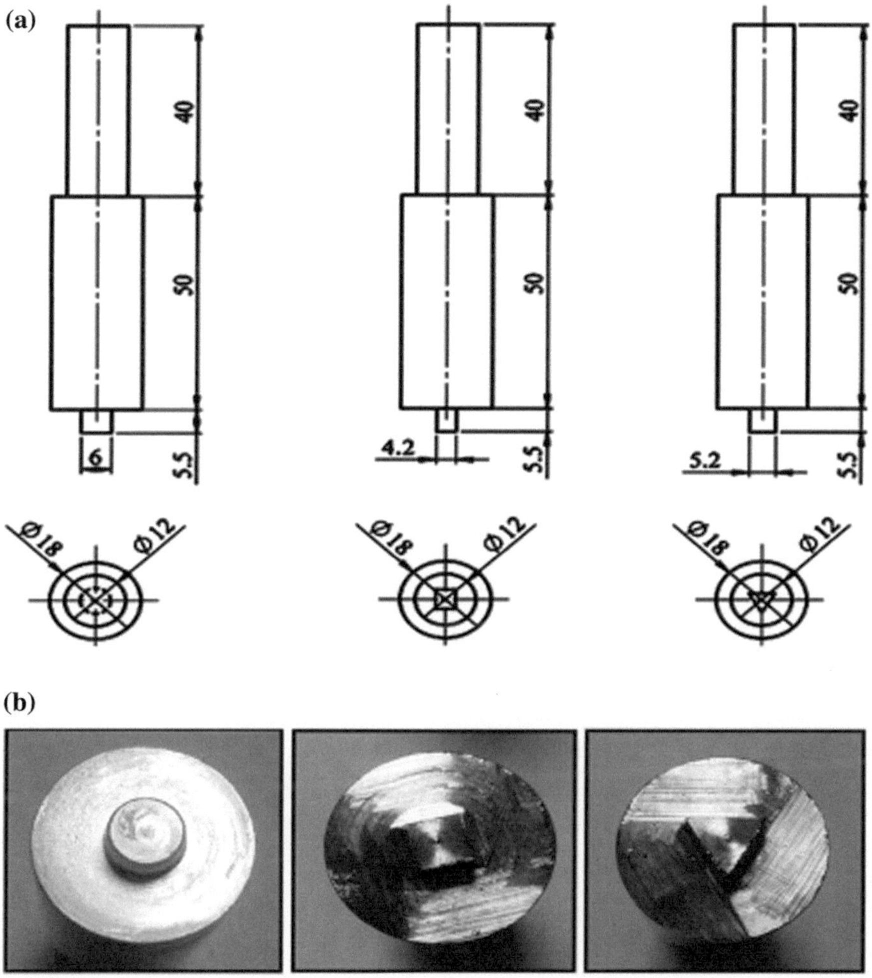

Fig. 1.4 **a** Dimensions of FSW tools and **b** photographic view of FSW tool pin profiles [12]

Area of application of FSW includes for the fabrication of parts for aerospace, automotive, electronic housings, coolers railway, shipbuilding, heat exchangers and nuclear waste containers. FSW has been in use in Europe since 1995 in production applications such as welding of extrusions, panelling for marine applications, rail car, heavy truck, medical applications, aluminium panels for deep freezing of fish on fishing boats, used to produce aluminium honeycomb panels and seawater resistant panels, prefabricated wide aluminium panels for high-speed ferryboats, aluminium tanks and boosters for spacecraft, for producing Ariane 5 motor thrust frames, for welding aluminium panels in the aircraft production, FSW of sandwich assemblies for a fighter aircraft fairing, friction stir welded hollow aluminium panels in the

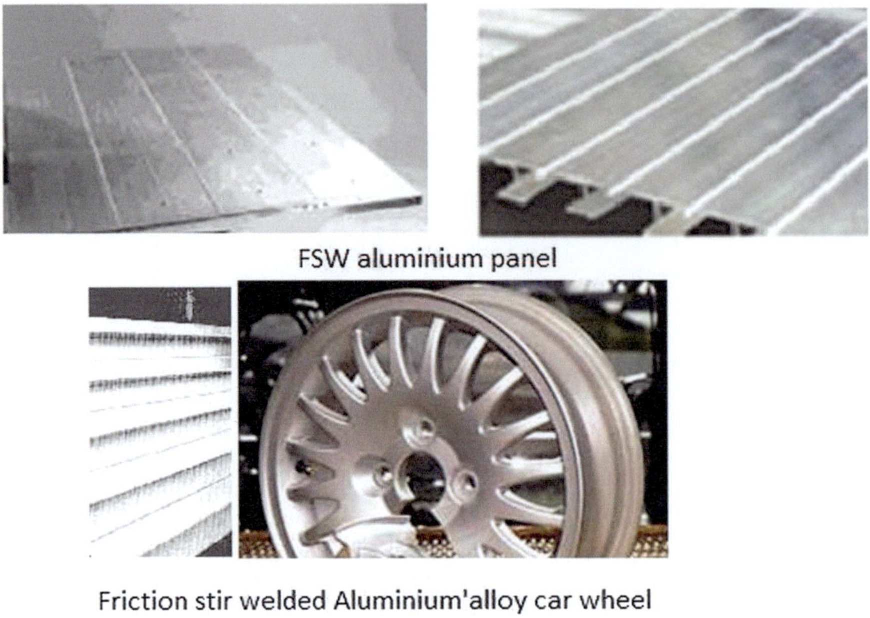

Fig. 1.5 Friction stir welded parts

rolling stock industry, friction stir welded aluminium components in the automotive industry, joining parts of a car wheel (see Fig. 1.5), friction stir welded hem joints and sandwich panels and friction stir welding of a wide range of workpieces as shown in Fig. 1.5.

1.4 Introduction to Friction Stir Processing

Friction stir processing is a variant of friction stir welding and very similar to friction stir welding in terms of experimental setup and tool, but the only difference is that, friction stir welding is for joining parts while friction stir processing is for microstructural modification and grain refinement for achieving improved material properties. Friction stir processing (FSP) is a microstructural modifications technique that is used to process material in solid state. FSP is an efficient manufacturing process for achieving homogeneous and refined grain structure in metal sheet. Friction stir processing is also an important technique in the field of superplasticity. FSP has greatly enhanced superplasticity in many Al alloys [27–39]. The principle of friction stir processing is based on the principle of friction stir welding. Friction stir processing has also been used to produce composite material especially in surface modification. In this section, friction stir processing is introduced with its principle of operation, advantages, limitations and areas of application.

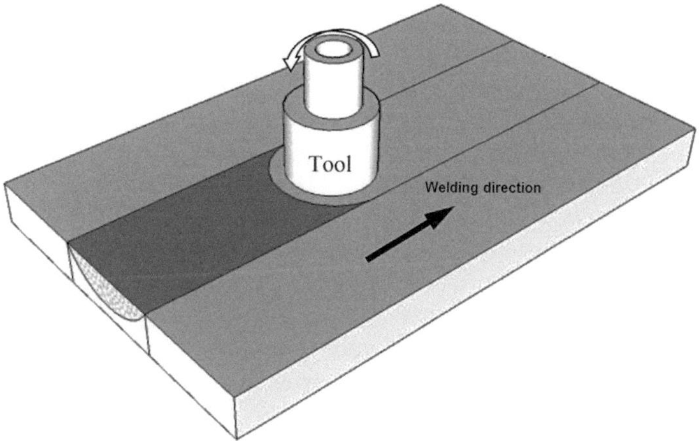

Fig. 1.6 Schematic diagram of friction stir processing

The principle of operation of friction stir processing is slightly different from that of the FSW in that the rotating tool penetration is limited because there is no need of joining materials. The rotating tool is plunged into the surface that needs to be processed and the tool travels through the surface to be processed. As the rotating tool progresses, the generated heat due to friction is used to plasticize the material around the processing zone and the grains are broken down through plastic deformation action of the rotating tool. In case of processing composite materials, the reinforcing material is introduced and the mixing of the reinforcing material with the matrix is achieved and further breaking down of grains and grain refinement continues as the FSP progresses. The schematic diagram of the process is shown in Fig. 1.6.

Friction stir processing produce highly refined, equiaxed grains that result in a number of improvement in important material properties such as strength, ductility and corrosion resistance. FSP has an advantage over other surface processing techniques because it is able to selectively refine the grains, as well as localized surface treatment, can be achieved. The FSP causes intense plastic deformation that produced a dynamically recrystallized fine, equiaxed, and defect-free grain structure. A number of high technology machines and equipment require materials with special properties. Aerospace and transportation industries are in constant need of high-performing materials to reduce their carbon footprint. The need to design materials with the desired properties is a major research problem that is constantly pursued in the research community. High strength and high ductility can be achieved with materials having small grain sizes. There are different processing techniques that can be used to achieve this objective, but friction stir processing is a new material processing technique that has outperformed the existing material processing techniques due to many attractive characteristics of the process. FSP helps to improve the formability of lightweight alloys at room temperatures that are needed in aerospace and automotive industries by refining and homogenizing the microstructure. The grain size is found to decrease causing the strain rate sensitivity to increase. The high difficulty in producing ultrafine grains in sheet metals that help to increase the superplasticity at lower temperature has been a stumbling block for the widespread utilization of lightweight alloys in the aerospace and automobile industries. The conventional grain refinement techniques based on thermo-mechanical processing such as hot rolling is very expensive, time consuming and not environmental friendly because of high energy consumption that is involved. Friction stir processing has been found to be an alternative material processing technique that can be used to produce ultrafine-grained microstructure and homogenized structure in one single step. The energy consumption in friction stir processing is very low and less time consuming. The tool used in FSP is simple and inexpensive and the process can readily be achieved on a milling machine. FSP can easily be automated, it is environmental friendly because no gases or chemicals are used, it has the ability to produce composite surface with high strength, high elastic modulus and high wear resistance [40], and the process is less expensive making it the process of choice over the other material processing techniques.

Some of the limitations of friction stir processing include limited data availability because the process is relatively new, lack of predictive models for predicting the resulting microstructure from the process and the keyhole at the end of each pass. All these limitations will be solved with more research in this field.

1.5 Summary

Friction welding, friction stir welding and friction stir processing, solid-state material processing techniques, were introduced in this chapter. Friction welding helps in joining bars of metal, pipes and other shape material in a cost-effective way. Friction stir welding has revolutionized the welding industries because of its ability to join material with thickness ranging between thin sections to a very large thickness with high weld quality. Friction stir processing, on the other hand, is based on the principle of friction stir welding process but it is used as a surface and bulk material modification technique that forces materials to flow at the temperature below the melting temperature of the material being processed. The tool action in this process causes extrusion of material and forged under high pressure to consolidate the material after cooling under hydrostatic pressure conditions. This material processing technique results in materials with high hardness, high strength and high ductility. FSP has also been used to modify the microstructure of reinforced metal matrix composite materials. The working principles of these processes are introduced in this chapter with the advantages, limitations and areas of application also highlighted in this chapter.

References

1. A.S. Bahrani, B. Crossland, *Friction Welding* (CME, 1976), pp. 61–66
2. F.D. Duffin, B. Crossland, Friction welding with sudden release of the fixed component, advances in welding processes, solid phase joining processes, in *Proceeding of the Conference*, The Welding Institute, Abington Hall, Cambridge (1971), pp. 25–33
3. B. Seshagirirao, V. Sivaramakrishna, G. Saikrishnaprasad, Experimental investigation of rotary friction welding parameters of Aluminum (H-30) and Mild Steel (Aisi-1040). Int. J. Innov. Res. Sci. Eng. Tech. **4**(5), 2920–2925 (2015)
4. W. Tomoyuki, U. Shigeharu, N. Tatsuyoshi, F. Masayuki, Properties of friction welds between 9Cr-ODS martensitic and ferritic–martensitic steels. J. Nucl. Mater. 367–370, 1213–1217 (2007)
5. P. Sammaiah, S. Arjula, G.R.N. Tagore, Mechanical properties of friction welded 6063 aluminum alloy and austenitic stainless steel. J. Mater. Sci. **45**, 5512–5521 (2010)
6. K. Jayabharath, M. Ashfaq, P. Venugopal, D.R.G. Achar, Investigations on the continuous drive friction welding of sintered powder metallurgical (P/M) steel and wrought copper parts. Mater. Sci. Eng. A 454–455, 114–123 (2007)
7. S. Mumin, Joining of stainless steel and copper materials with friction welding. Ind. Lubr. Tribol. **61**, 319–324 (2009)
8. C. Shanjeevi, S. Satish Kumar, P. Sathiya, Evaluation of mechanical and metallurgical properties of dissimilar materials by friction welding. Procedia Eng. **64**, 1514–1523 (2013)
9. W. Radosław, K. Mieczysław, Friction welding of ductile iron with stainless steel. J. Mater. Process. Tech. **213,** 453–462 (March 2013)
10. C.G. Subhash, R.G. Madhusudhan, R.A. Venugopal, Influence of rotational speed on microstructure and mechanical properties of dissimilar metal AISI 304-AISI 4140 continuous drive friction welds. J. Iron Steel Res. Int. **19**, 64–73 (2012)
11. H. Jian, L. Huijun, Z. Zhixiong, B. Frank, J. Laizhu, X. Haigang, M. Li, Microstructure and mechanical properties of friction stir welded 18Cr–2Mo ferritic stainless steel thick plate. Mater. Des. **63**, 238–246 (November 2014)
12. S. Senthil Kumar, S. Narayanan, S. Denis Ashok, Acoustic emission–based monitoring approach for friction stir welding of aluminum alloy AA6063-T6 with different tool pin profiles. Proc. Inst. Mech. Eng. Part B J. Eng. Manuf. **227**, 407–416 (2013)
13. M. Jafarzadegan, A. Abdollah-zadeh, A.H. Feng, T. Saeid, J. Shen, H. Assadi, Microstructure and mechanical properties of a dissimilar friction stir weld between austenitic stainless steel and low carbon steel. J. Mater. Sci. Tech. **29**, 367–372 (April 2013)
14. E.T. Akinlabi, S.A. Akinlabi, Effect of heat input on the electrical resistivity of dissimilar friction stir welded joints of aluminium and copper, in *IEEE Africon International Conference*, Livingstone, Zambia, 13–15 September 2011
15. B. Sadeghian, A. Taherizadeh, M. Atapour, Simulation of weld morphology during friction stir welding of aluminium stainless steel joint. J. Mater. Process. Technol. **259**, 96–108 (2018)
16. I.J. Ibrahim, G.G. Yapici, Application of a novel friction stir spot welding process on dissimilar aluminum joints. J. Manuf. Processes **35**, 282–288 (2018)
17. M.-J. Hsieh, R.-T. Lee, Y.-C. Chiou, Friction stir spot fusion welding of low-carbon steel to aluminum alloy. J. Mater. Process. Technol. **240**, 118–125 (2017)
18. S. Honggang Dong, Y.S. Chen, X. Guo, X. Zhang, Z. Sun, Refilled friction stir spot welding of aluminum alloy to galvanized steel sheets. Mater. Des. **94**, 457–466 (2016)
19. R. Ranjana, A.C. de Oliveira Mirandab, S.H. Guoa, S. Walbridgea, A. Gerlichc, Fatigue analysis of friction stir welded butt joints under bending and tension load. Eng. Fract. Mech. **206**, 34–45 (2019)
20. B. Sadeghian, A. Taherizadeh, M. Atapour, Simulation of weld morphology during friction stir welding of aluminum stainless steel joint. J. Mater. Process. Technol. **259**, 96–108 (2018)
21. W.M. Thomas, Friction stir butt welding. US Patent No. 5,460,317, USA (1991)
22. N.T. Kumbhar, K. Bhanumurthy, Friction stir welding of Al 6061 alloy. Asian J. Exp. Sci. **22**(2), 63–74 (2008)

23. M.K. Kulekci, E. Kalucx, S. Aydınk et al., Experimental comparison of MIG and friction stir welding processes for EN AW-6061-T6 aluminium alloy. Arab. J. Sci. Eng. **35**(1B), 321–330 (2010)
24. K. Elangaovan, V. Balasubramanian, Influences of pin profile and rotational speed of the tool on the formation of friction stir processing zone in AA2219 aluminum alloy. J. Mater. Sci. Eng. **459**(1–2), 7–18 (2007)
25. P. Bahemmat, A. Rahbari, M. Haghpanahi et al., Experimental study on the effect of rotational speed and tool pin profile on AA2024 aluminum friction stir welded butt joints, in *Proceedings of ECTC 2008, ASME Early Career Technical Conference*, Miami, FL, 3–4 October 2008, pp. 1.1–1.7
26. H.S. Patil, S.N. Soman, Experimental study on the effect of welding speed and tool pin profiles on AA6082-aluminum friction stir welded butt joints. Int. J. Eng. Sci. Technol. **2**(5), 268–275 (2010)
27. M. St. Węglowski, A. Pietras, Friction stir processing—analysis of the process. Arch. Metall. **56**(3), 779–788 (2011)
28. R. Palanivel, M.P. Koshy, The tensile behaviour of friction-stir welded dissimilar aluminium alloys. Mater. Technol. **45**, 623–626 (2011)
29. W.B. Lee, Y.-M. Yeon, S.-B. Jung, Mechanical properties related to micro structural variation of 6061 Al alloy joints by friction stir welding. Mater. Trans. **45**(5),1700–1705 (2004)
30. X. Qi, Z. Hao, D. Wildermuth, In-process tool monitoring through acoustic emission sensing. SIMTech Technical Report (AT/01/014/AMP), Singapore Institute of Manufacturing Technology, pp. 1–8 (2001)
31. C.N. Suresha, B.M. Rajaprakash, S. Upadhya, Applicability of acoustic emission in the analysis of friction stir welded joints. Int. J. Recent Trends Eng. **1**(5), 86–89 (2009)
32. S.-K. Oh, A. Hasui, T. Kunio et al., Effects of initial energy on acoustic emission relating to weld strength in friction welding. Trans. Jpn. Weld Soc. **13**(2), 15–26 (1982)
33. H.-K. Yoona, Y.-S. Kongb, S.-J. Kimb et al., Mechanical properties of friction welds of RAFs (JLF-1) to SUS304 steels as measured by the acoustic emission technique, in *Proceedings of the Seventh International Symposium on Fusion Nuclear Technology—ISFNT-7 Part B*. Fusion Eng. Des. **81**(8–14), 945–950 (2006)
34. C. Chen, R. Kovacevic, D. Jandgric, Wavelet transform analysis of acoustic emission in monitoring friction stir welding of 6061 aluminum. Int. J. Mach. Tool Manuf. **43**, 1383–1390 (2003)
35. M. Karthikeyan, A.K. Shaik Dawood, Influence of tool design on the mechanical properties and microstructure in friction stir welding of AA6351 aluminium alloy. IRACST—Eng. Sci. Technol.: Int. J. **2**(2), 233–237 (2012)
36. V. Soundararajan, M. Valant, R. Kovacevic, An overview of R&D work in friction stir welding at SMU. Metalurgija **12**(204), 277–295 (2006)
37. R. Mishra, L. Johannes, I. Charit, A. Dutta, Multi-pass friction stir superplasticity in aluminum alloys. Proc. NSF 931–935 (2005)
38. Z. Ma, R. Mishra, M. Mahoney, Superplasticity in cast A356 induced via friction stir processing. Scripta Materialia **50**, 931–935 (2004)
39. A. Dutta, I. Charit, L. Johannes, R. Mishra, Deep cup forming by superplastic punch stretching of friction stir processed 7075 Al alloy. Mater. Sci. Eng., A **395**, 173–179 (2005)
40. E.T. Akinlabi, R.M. Mahamood, S.A. Akinlabi, E. Ogunmuyiwa, Processing parameters influence on wear resistance behaviour of friction stir processed Al-TiC composites. Adv. Mater. Sci. Eng. **2014** (2014). http://www.hindawi.com/journals/amse/2014/724590/

Chapter 2
Friction Welding

2.1 Introduction

Friction Welding is a solid-state welding process, where there is no application of heat from any external source but the heat required for the welding process to be actualized is generated from friction that is created by the relative motion of the surfaces to be joined. The heat generated is high enough to bring the mating surfaces into what is called the plastic stage and with the application of required pressure, the two surfaces coalesced. This is a solid-state welding process because there is no melting of the workpiece involved during the entire welding process [1–6]. The relative movement of the workpiece to be joined results in friction that displaces material plastically on the contact surfaces and high pressure is applied until the welding process is completed. This welding process was developed by the Soviet Union in 1956 while the machines for this welding process were developed by American companies: Caterpillar, Rockwell International and American Manufacturing Foundry [5].

2.1.1 Working Principle of Friction Welding Process

There are many types of friction welding processes that work differently, but all the different friction welding processes involve a common working principle as outlined below:

- The weld is produced under high compressive pressure forces and the material being welded remains in the solid-state throughout the entire welding process.
- Before the commencement of the friction welding process, the workpieces need to be prepared so that the two surfaces are flat, smooth, and square-cut surfaces. This is required to have complete welding of the required surfaces otherwise partial welding will occur only on the contact areas of the two faces.

E. T. Akinlabi and R. M. Mahamood, *Solid-State Welding: Friction and Friction Stir Welding Processes*, Mechanical Engineering Series,
https://doi.org/10.1007/978-3-030-37015-2_2

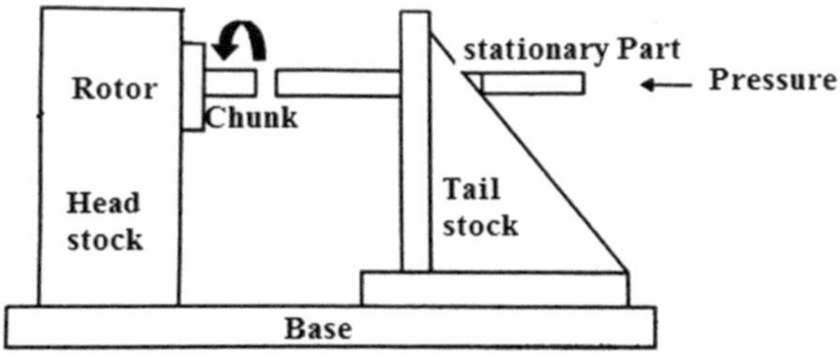

Fig. 2.1 Schematic diagram of friction welding

- One of the workpieces is mounted on a chuck that is attached to a rotor while the other workpiece is held stationarily. To achieve the required surface flatness of the two parts to be joined, a little pressure force is applied on the stationary workpiece to come in contact with the rotating workpiece which allows the cleaning and flattening of the two surfaces through the burnishing action.
- The workpiece that is mounted on the rotor is then rotated at high speed against the stationary workpiece and the rotating part is brought in contact with the stationary part by the high pressure that is applied to the stationary workpiece forcing it against the rotating workpiece. The contact of the workpieces rotating relative to one another under high pressure generates heat as a result of the high friction force between the two surfaces. The process continues until the generated heat reaches the required welding temperature that is used to plastically deformed and displace the material from faying surfaces and form as flash. The flash can subsequently be removed in a finishing operation if desired.
- When the desired welding temperature is reached, then the rotor is stopped. The applied pressure is then increased until the desired time elapses and the welding process is complete. Upon colling, the two parts are seen to have joined with very strong weldment produced at the joint interface. The schematic diagram of friction welding process is shown in Fig. 2.1 and the schematic diagram of the different stages involved in friction welding process is shown in Fig. 2.2.

The weld produced is a solid-state fusion of the two components that is cut across the entire face of the joint and the weld integrity is of superior quality. The different types of friction welding are presented in the next subsection.

2.2 Types of Friction Welding Process

There are different variants of Friction welding that all work on the same principle. They are: rotary/spin friction welding consisting of continuous friction welding and inertial induce friction welding, linear friction welding, linear vibration welding and friction surfacing. Each of these friction welding processes is described in the following subsection.

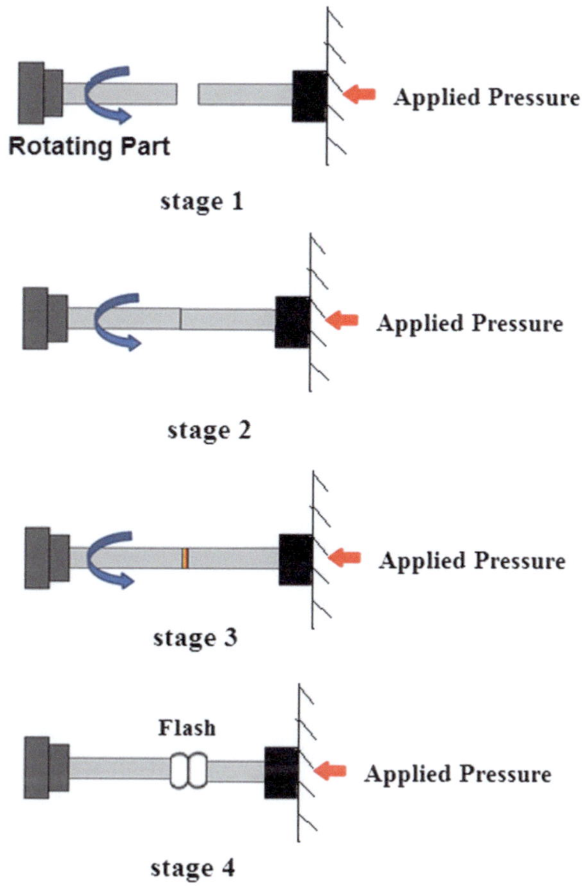

Fig. 2.2 Stages of friction welding process

2.2.1 Rotary/Spin Friction Welding

The rotary friction welding also known as spin friction welding. It is the type of friction welding process that was described above that involves rotation of one part against the other stationary parts. There are two main types of rotary friction welding processes, namely continuous induced friction welding and inertia friction welding. These friction welding processes are explained in the following subsections.

2.2.1.1 Continuous Induce Friction Welding

Continuous induce friction welding is the one in which the rotor is connected with a band brake. The band brake is used to stop the rotor when the desired welding

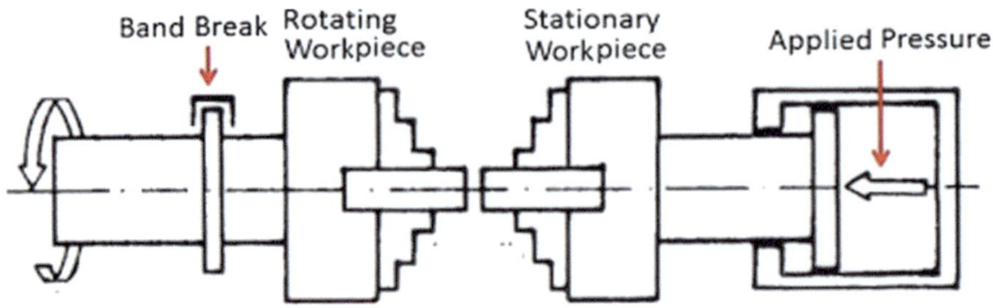

Fig. 2.3 Schematic diagram of continuous friction welding process

temperature is reached and the pressure is then continued to be applied on the workpiece until the weld is formed. The schematic diagram of continuous indiced friction welding is shown in Fig. 2.3.

2.2.1.2 Inertia Friction Welding

Inertial friction welding is another variant of friction welding and the only modification in this variant is the replacement of the band brake that is used to stop the rotational motion in the continuous friction welding process, with two flywheels. One of the two flywheels is the engine flywheel and the other one is the shaft flywheel [6]. The two flywheels are connected together at the beginning of the welding process and the flywheels are used to connect the chuck to the motor as shown in Fig. 2.4. When the desired speed is attained, the engine flywheel is disengaged from the shaft flywheel. As the two flywheels are separated, the shaft flywheel will later stop automatically without the need for a brake due to the low moment of inertia of this flywheel, while the pressure force is continuously applied to the workpiece until the weld is formed.

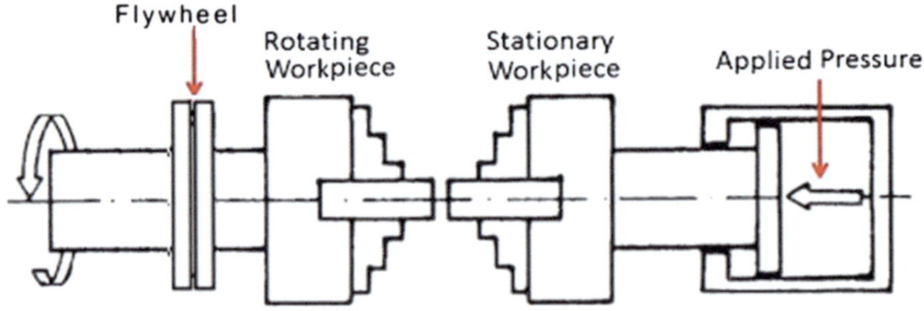

Fig. 2.4 Schematic diagram of inertial friction welding process

2.2.2 Linear Friction Welding

Linear friction welding process is similar to the rotary/spin friction welding process but the difference is that the chuck rotates circularly in spin friction welding process, while the chuck oscillates in the linear friction welding process instead of spinning. Also, the operating speed in the linear friction welding process is lower than in the spin welding process [7]. The quality of weld produced in the linear friction welding process is higher than that obtained in the spin friction welding process; the high-quality weld is attributed to the low speed in the linear friction welding process. The schematic diagram of the linear friction welding process is shown in Fig. 2.5. The linear welding process is achieved by keeping one of the parts to be welded stationary while the second part is moved in a reciprocating manner and pressing it against the part that is held stationary as shown in Fig. 2.5. The friction between the two parts in relative motion and under a high compressive force generates heat that causes the surfaces of the two parts to be plasticized. As the process continues, the plasticized material is forced out from the interface as flash that causes the parts to be reduced in length in the direction of the applied compressive force while the desired bond is

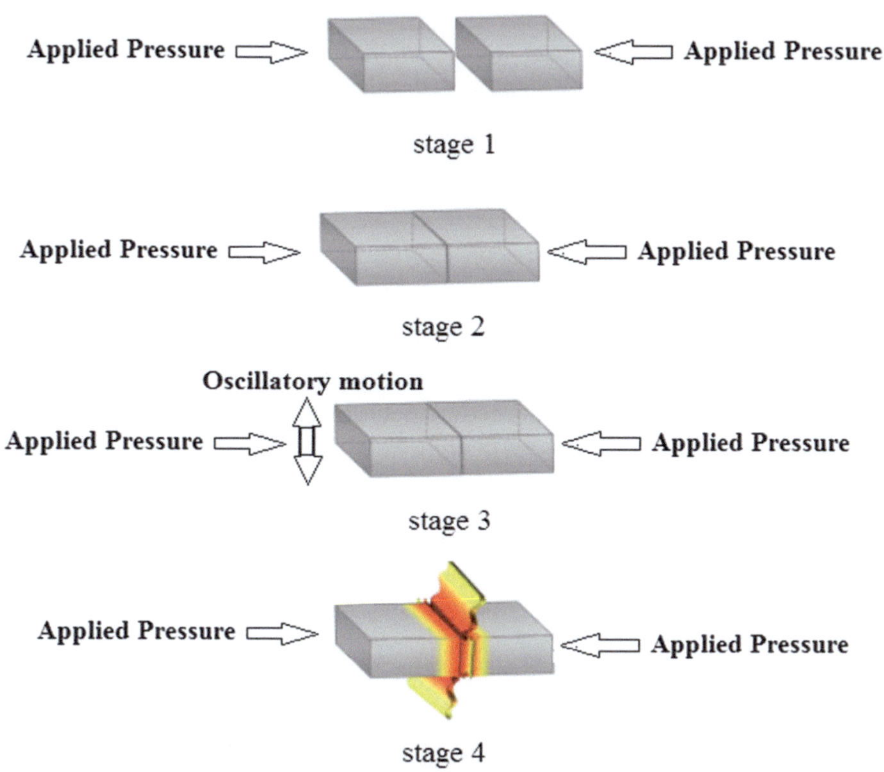

Fig. 2.5 Schematic diagram of linear friction welding process

achieved. Linear Friction Welding is an important type of friction welding process that is used by manufacturers in many industries for the fabrication process of parts with various geometric shapes unlike only circular parts that are permitted in the spin friction welding process. It gives superior weld quality even more than the traditional fusion based welding processes. It has a narrow heat-affected zone with fine grain structure and a smooth transition from the weld zone to the parent material. The process can be used to join any shape of part no matter the complexity. It can also be used to join dissimilar materials for the fabrication of bimetallic parts with near-net shapes. This type of friction welding process also saves manufacturing time because it is faster and many parts can be joined together simultaneously thereby reducing the manufacturing cycle time.

Linear friction welding process does not require large preparation time which helps to reduce the overall cycle time and help to increase productivity. The process is repeatable and can easily be automated.

2.2.3 *Vibration Welding*

Vibration welding is another type of friction welding process that is basically used in welding thermoplastic parts with planar or slightly curved surfaces. In the vibration welding process, the welding is achieved by placing materials in contact with one another and high pressure is applied [8]. Then an external vibration force is applied in order to rub the pieces against one another to provide friction and generate heat at the interface of the two materials to be joined. The generated heat causes the melting of the materials at the interface and the molten materials are forced to flow and mix in order to form a strong bond with the application of pressure, and upon cooling a strong weld is produced. Vibration welding is of two types, namely linear and orbital vibration welding. Linear vibration welding uses oscillatory or back-and-forth motion of one of the workpieces to produce friction that generates the heat required for the welding process. A circular motion is used in vibrating the upper part of the workpiece to produce friction and generate the needed heat for the welding in the orbital friction welding process. In orbital friction welding process, under axial pressure, one of the workpieces is used to rub the surface of the other workpieces in a circular motion throughout the peripheral of the interface at all points around the perimeter that constantly changes from transverse motion to longitudinal motion. The vibration welding process is made up of four stages that are similar to the one described above. The first stage is when the vibration causes the temperature at the interface to be increasing up to when the glass transition temperature of amorphous thermoplastics or the crystalline melting point of semicrystalline plastics is reached. This is followed by the second stage when the material at the interface begins to melt and the flow of the molten material causes deeper weld penetration. The third stage is when the weld penetration continued to increase with time while the melting and viscous flow attain a steady state and the vibration of the workpiece is stopped. The last stage sees more weld penetration as pressure is further applied and the molten

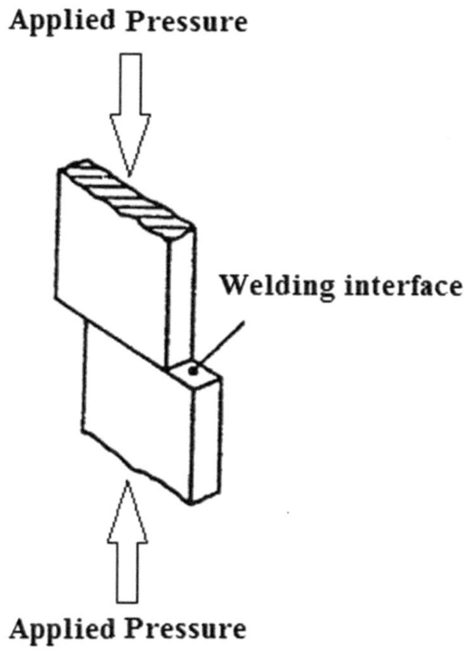

Fig. 2.6 Linear vibration welding

materials solidify and resulting in the formation of strong weld. The orbital vibration welding is applicable when joining plastic parts made of irregularly shapes while the linear vibration welding is mostly used for joining regular-shaped plastic parts. The schematic diagram of the linear vibration welding process is shown in Fig. 2.6.

Vibration welding operates at a frequency of between 120 and 240 Hz frequencies, welding pressure of 0.5–20 MPa and a weld time of about 5–15 s or 4–10 cycles per minute

2.2.4 Friction Surfacing

Friction surfacing process is a coating process for improving the surface properties such as wear resistance and corrosion resistance properties of the substrate material [9]. This coating process is achieved by rotating the coating material under pressure, against the substrate material, as the substrate material is moved in lateral motion against the rotating material, the plasticized rotating material is seen deposited on the path of the moving substrate. The coating material is consumable in this process and it kept decreasing as the process is progressed. The coating material is referred to as 'mechtrode'. The schematic diagram of the friction surfacing process is shown in Fig. 2.7. A pictorial diagram of friction surfacing of aluminium on mild steel plate is shown in Fig. 2.8 [10].

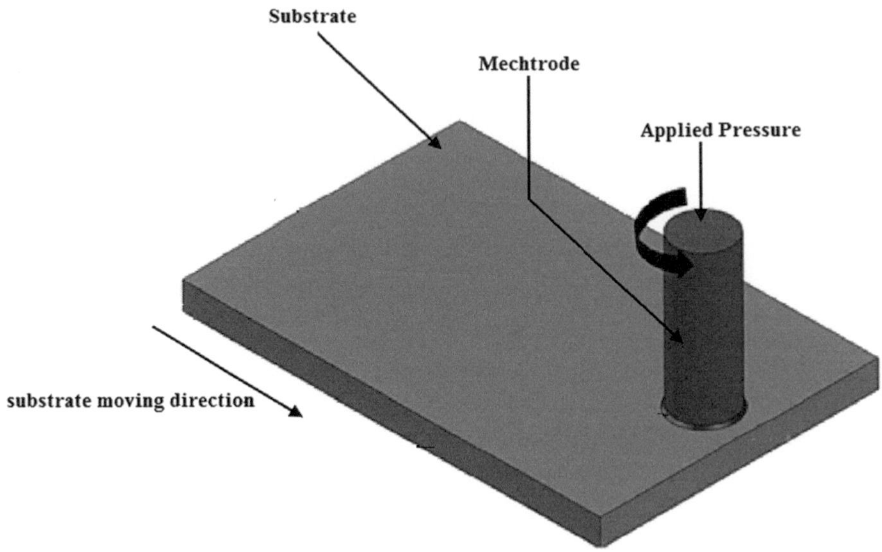

Fig. 2.7 Schematic diagram of friction surfacing process

Fig. 2.8 Coating of Aluminium over the Mild Steel plate [10]

Friction welding and its variants are very important manufacturing processes and have found their applications in different areas. The advantages, limitations and areas of applications of friction welding process are presented in the next subsection.

2.3 Advantages, Limitations and Areas of Application of Friction Welding

Friction welding is very useful to produce solid-state welding and comes with lots of advantages. Some of these advantages are as follows:

- Some of the problems encountered in fusion welding processes are offset with the friction welding process that includes porosity, crack formation and part distortion.

The friction welding is achieved in solid state, and it does not involve melting thereby eliminating the aforementioned problems and the distortion of the welded part is greatly reduced.
- Friction welding is an environmental-friendly process and it does not generate smoke or any greenhouse gases.
- In the friction welding process, the peak temperature is considerably lower than the peak temperatures obtainable in fusion welding which helps to reduce the formation of intermetallic compounds and hence allows joining of various dissimilar materials.
- The quality of weld produced using friction welding is comparable and sometimes more superior in strength to the parent material, and the fatigue performance of the weld produced in some materials have been found to be higher.
- The heat-affected zone produced in friction welding is smaller or narrower when compared to fusion welding making it a choice welding process for heat-sensitive materials.
- Friction welding does not require the use of filler material and the use of flux and shielding gas is not needed, making the process to be rapid and cheaper.
- The process is highly repeatable and less human dependant that makes the defect rates to be lower and the process can thus be easily automated.
- Friction welding helps to improve the buy-to-fly ratio because it reduces the quantity of material needed to make a component. Workpiece of any shape and size can be welded to produce a preform that can subsequently be machined which significantly reduces manufacturing time and costs.
- Friction welding is an energy-efficient process and fast manufacturing process. It is an important welding technology that helps to increase the material service life and for product remanufacturing.
- The efficiency of this welding process is very high.

Some of the limitations of the friction welding process include the following:

- Materials that cannot be forged cannot be welded.
- Only angular and flat butt joints can be produced.
- Cost of the setup of machine is very high.
- It can only be used for smaller parts of machines and it is not applicable to thin metal.

Friction welding is useful in many applications, some of these are listed below:

- Friction welding and its variants can be used to join metals and thermoplastics in the construction of different types of components and parts in the aerospace and automotive industries. Vibration welding is mostly used in automotive assembly applications that include intake manifolds, tail lights and lenses, door panels, fluid reservoirs and bumpers. Also used in aerospace application such as in interior lighting, ducts and overhead storage bins. It is also used in consumer products such as dishwasher pumps, soap dispensers, vacuum cleaner housings, toner cartridges, display stands and shelves and hospital bedpans.

- Linear friction welding is used in the fabrication of integrated bladed disks (blisks) in aero engines. This process helps to significantly reduce the weight of blisks when compared to those produced using conventional manufacturing processes. Also, the removal of the mechanical interface between the blades and the disk help to eliminate the stress riser site where fatigue crack is usually initiated. Friction-welded blisks give better aerodynamic performance.
- Friction welding can be used to join shafts that cannot be produced with other welding processes. Other components such as pipes, gear and valves can also be easily welded using friction welding for many applications including marine and oil industries.
- Friction welding is used to replace forging or casting assembly.
- The use of friction welding to join dissimilar materials such as copper and aluminium found their applications in industries including electrical and electronic industries.
- Friction welding and its variants find a wide range of applications in many industries including for the manufacturing of structural components and in the nuclear power plant.

2.4 Processing Parameters in Friction Welding

The most important process parameters in the friction welding and its variants are the friction pressure—the pressure that is required to generate friction and the required heat for the welding, the forging pressure or upset pressure—this is the axial pressure that is applied on the stationary workpiece once the movement is stopped, the burn-off length—this is the overall length lost after welding), welding speed—rotational speed, translational speed, etc. and welding time. The friction pressure is the pressure that is maintained between the parts to be welded during the rotational or oscillational motion period. All these processing parameters have a great influence on the properties and performance of weld produced. The forging pressure is the pressure also between the parts to be joined that is usually kept for a short period of time after the movement is stopped and is typically greater than the friction pressure. The strength of the weld is dependent largely on the forging pressure. The burn-off is the linear measurement of the material that is transformed into "flash". Flash is the material that is forced out of the two contacting surfaces during friction welding process and it is seen around the welding. The rotational speed in friction welding is another important welding parameter and it influences the quality of weld that is produced. Too low welding speed will not generate sufficient quantity of heat required for the welding process while too high speed will produce a wide heat-affected zone as well as a large quantity of flash. The speed and all processing parameters should be carefully controlled in order to achieve the desired weld properties. The rotational speed, for example, usually depends on the workpiece material and the diameter to be welded. The weld time is another process parameter that needs to be optimized to achieve desired weld properties and performance. The weld time depends on materials to be welded. It also depends on the shape of the weld and surface area of the workpiece. A number of research works have appeared in the literature on the study of processing parameters of friction welding process. Some of these research works are presented in the next section.

2.5 Research Advancement in Friction Welding

A number of research works are reported in the academic literature on friction welding process, and the process parameter governing the welding process. Ravi et al. [11] investigated the influence of processing parameters—friction pressure, upset pressure, burn-off and rotational speed on metallurgy and mechanical property of friction-welded hollow head and a solid stem of an engine valve using martensitic stainless steel (X45CrSi93) of 6.5 mm diameter and 3 mm of inner diameter of the hollow head (see Fig. 2.9). The results showed that the friction pressure is the most significant process parameter with a P-value of 0.02 and it has a greater influence on the tensile strength. The upset pressure is also significant with a P-value of 0.033 and the rotational speed has a P-value of 0.036. The authors used a response surface methodology to determine optimum welding parameters and the regression equation was used to predict the maximum tensile strength for the optimal parameters. The optimal process parameters as obtained from the experimental results are 17.2 MPa of friction pressure, 31 MPa of upset pressure, 11 mm of burn-off length and 1800 rpm

Fig. 2.9 Friction welded samples [11]

of spindle speed which gives the maximum tensile strength value of 891.552 MPa. The experimental result is in good agreement with the predicted values.

In a similar study, Sahin [12] investigated the process parameters influence on the mechanical properties of the samples produced using friction welding of dissimilar materials of stainless steel and copper. The author developed an empirical model using a response surface methodology to predict the strength of the joints. The results showed that the friction pressure/friction time has the largest influence on the tensile strength of the joints. The rotational speed was also found to significantly influence the tensile strength of the joint. Higher hardness was found to be caused by the formation of intermetallic compound ($FeCu_4$ and Cu_2NiZn) at the interface. The optimized process parameters were found to produce a sample with a maximum tensile strength of 223 MPa at a friction pressure/time relation of 8.82 MPa/s, with an upset pressure/time 8 MPa/s and a rotational speed of 23.5 revolutions per second. The heat-affected zone was also found to be small as shown in Fig. 2.10.

Sreenivasan et al. [13] studied the influence of process parameters of rotary friction welding of Al–SiC composite on the mechanical properties of the friction-welded joints. The process parameters that were studied are the spindle speed, the friction pressure, the upset pressure and the burn-off-length. The influence of these processing parameters on the ultimate tensile strength and hardness was investigated and the optimized process parameters were obtained using Genetic Algorithm (GA). The predicted optimized process parameters were validated with experimental data and a very close agreement was obtained. The optimized process parameters obtained are: spindle speed of 1491.54 rpm, friction pressure of 98.94 MPa, upset pressure of 209.26 MPa, burn-off length of 1.5 mm to give the maximum ultimate tensile strength of 244.2866 MPa and hardness of 148.2392. Titouche et al. [14] investigated the

Fig. 2.10 Image of the interface of a joint [12]

mechanical and electrochemical behaviour of the friction-welded titanium stabilized austenitic stainless steel (AISI 321) thick tube in a nuclear reactor. Four heat diffusion times' influence on the mechanical properties (tensile strength and microhardness) of the sample produced was investigated. The results revealed fine grain structure at the weld zone as well as deformation-induced alpha martensitic microstructure that increases from the parent metal towards the centre of the weld with a quasi-Gaussian hardness profile that was attributed to the α'-martensitic transformation and grain refinement in the weld zone. The fracture site of the tensile specimen was found to be in the heat-affected zone and base metal and not at the weld zone or the hard recrystallized and the thermo-mechanical affected zone. The samples that had fractures occurring at the weld zone are samples with poor weld quality and produced with insufficient heat input. Also, the ductility of the weld was found to increase as the friction time was increased and then decrease as the friction time was increased beyond 8.5 s. The corrosion resistance of the central-welded zone was found to be improved when compared to the heat-affected zone and the parent metal. Khidhir and Baban [15] also investigated the effects of frictional welding processing parameter (forging pressure) on the microstructure and mechanical properties of friction welded AISI 1045 medium carbon steel and AISI 316L austenitic stainless steel. All other processing parameters: the friction pressure, friction time, forging time and rotational speed were kept constant. The result showed that the hardness of the weld interface is harder than the parent material and it increases as the forging pressure increases, while the tensile strength was found to decrease with increasing forging pressure. The welded sample fracture in the thermo-mechanical-affected zone (TMAZ) on the 316L austenite stainless steel side during tensile testing. The microstructure showed that the widths of the TMAZ and the heat-affected zone (HAZ) were larger in the AISI 1045 medium carbon steel than those of the 316L ASS.3 and it was concluded that the forging pressure has a significant influence on the ultimate tensile strength of the weld joint. Linear friction welding of two titanium alloys grade five was conducted by Homma et al. [16] and the microstructural analysis was performed. The microstructures in the weld zone were found to be deformed, while there are locally built up weak strains seen in the TMAZ as a result of plastic deformation. Also, a colony of fine twinning, α', α and β phases were found in the weld interface and the morphology of the twinning is a deformed needle-like structure, while in the TMAZ, locally piled up weak strains remain due to the plastic deformation. The weak local strains are accommodated in the HAZ probably due to the heat effect. Cheniti et al. [17] also conducted a rotary friction welding of WC–Co cermet to AISI 304L austenitic stainless steel by varying the friction times and performed microstructural analysis and mechanical properties evaluation. The physical examination of the samples showed that the flash length increases as the friction time increases as shown in Fig. 2.11. The weld joints produced from 4 and 6 s friction times show the absence of flash while the flash length observed for 8 s friction time is very small and increases as the friction time is increased. The microstructural evaluation showed that the grain sizes in the heat-affected zone and the thermo-mechanically affected zone increased as the friction time was increased. A Fe–Cr–W-rich band was seen along the WC–Co/AISI 304L weld interface in the

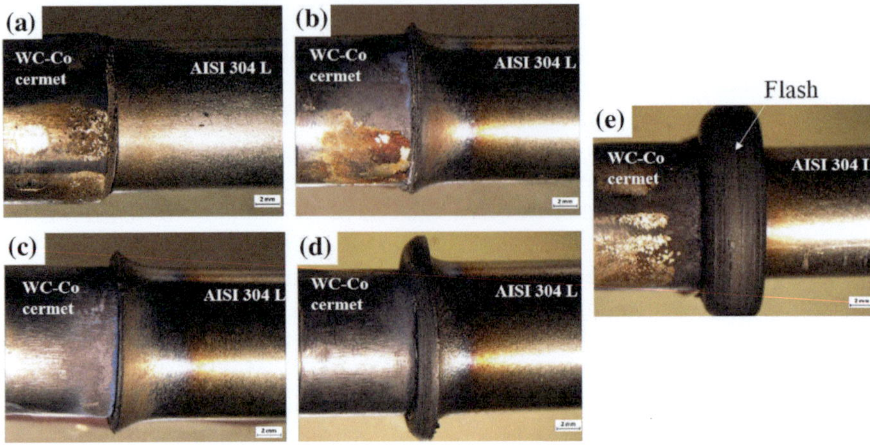

Fig. 2.11 Photos of the weld joints obtained using different friction times, **a** 4 s, **b** 6 s, **c** 8 s, **d** 10 s and **e** 12 s [17]

central region of the weld that was found to increase with increasing friction time. The band formation was attributed to the occurrence of interdiffusion between the cermet and the steel that enhanced the metallurgical bonding of the weld interface. The mechanical behaviour showed that increasing the friction time causes an increase in hardness and Young's modulus of the weld zone.

Pissanti et al. [18] investigated the microstructure of the weld cross section and the Charpy impact toughness at 0 °C for the different regions in a pipeline girth friction welding of the UNS S32205 alloy. The results showed that there is a great deal of microstructural modifications at the weld interface as a result of the very high temperature generated by the friction as well as the plastic deformation has taken place during the friction welding process. Also, there is a considerable drop in impact toughness at the weld interface due to substantial changes in austenite morphology and excessive ferrite grain growth.

Jedrasiak et al. [19] proposed a finite element thermal model for linear friction welding of Ti6Al4V. The developed model captured the lateral heat input distribution at the interface, the geometrical changes and heat loss due to the expulsion of flash using a sequential step-wise technique.

The authors developed a two-dimensional thermal model for the linear friction welding of Ti6Al4V. The temperature predictions by the model are in good agreement with the experimental data and it is computationally efficient. Stütz et al. [20] investigated inertial friction welding of tubular components of pure molybdenum of outside diameter of 150 mm, and inside diameter of 130 mm, using the existing inertial friction welding process parameters for TZM tubes. Compared with TZM tubes, the welding of molybdenum was more challenging with high upset rates and high motor load during the friction stage of the friction welding process. The welds produced are defect free and with fine-grained microstructure. The extensive plastic deformation of the weld zone was attributed to the higher thermal diffusivity and the lower strength of the molybdenum as compared to TZM (see Fig. 2.12).

The contact zones of the Mo and TZM as shown in Fig. 2.12b and c showed a fine-grained microstructure that is produced as a result of severe plastic deformation and recrystallization in the weld zone. The increase in upset was found to increase with a decrease in spindle speed and for Mo, a low force gradient make a successful weld realizable.

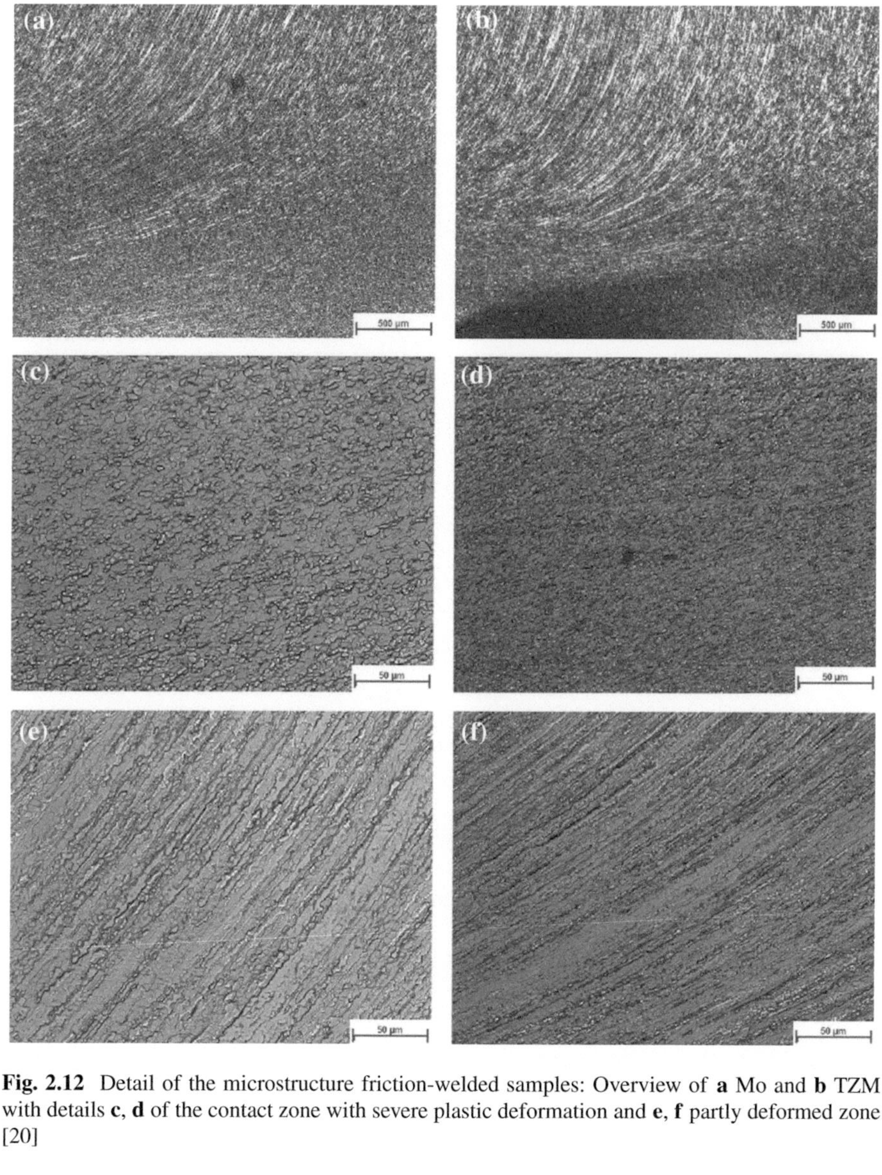

Fig. 2.12 Detail of the microstructure friction-welded samples: Overview of **a** Mo and **b** TZM with details **c**, **d** of the contact zone with severe plastic deformation and **e**, **f** partly deformed zone [20]

The formation of intermetallic compound when joining dissimilar materials such as titanium and steel using the fusion welding technique is a challenge that influences the property of the weld. Friction welding has helped to reduce the formation of intermetallic compound with proper control of processing parameters. Cheepu et al. [21] investigated the bonding interface of friction-welded dissimilar material of titanium and stainless steel using a new technique of electrodeposition of nickel coating on stainless steel as an interlayer. The result showed that the tensile strength of the nickel interlayer joints was higher than that of the joint without interlayer. The interface of titanium and stainless steel with the interlayer showed the presence of Ti–Ni phases which is more plastic than the brittle Fe–Ti intermetallic compounds that is formed without the interlayer. The study showed that the friction welding processes using interlayers will be an attractive route for joining various combinations of dissimilar materials due to their poor weld properties. A sound weld between Ti and 304 SS using the Ni interlayer can be produced using the right combination of process parameters. The processing parameters were found to have a great influence on the properties of the interlayer. By increasing the upsetting pressure and the heating time resulted in the reduction of interlayer thickness and an increase in strength of the joints within the optimal conditions. The weld flash formation at the joint interface of the welds at different upset pressure are shown in Fig. 2.13. The cross-sectional view of the joints showed that the titanium underwent greater deformation than the stainless steel, and axial shortening is more in titanium while that of the stainless steel is not low. The use of higher upset pressure and friction time increased the weld length of weld flash and caused axial shortening. The upset pressure was varied between 130 and 180 MPa, P1 = 130 MPa, P2 = 140 MPa, P3 = 150 MPa, P4 = 160 MPa, P5 = 170 MPa and P6 = 180 MPa. The amount of weld flash gradually increased from half radius to a complete ring from low upset pressure to higher upset pressures. The low amount of flash was extruded from the interface at low-pressure P1 with a large width of the unmixed white region at the interface as seen in Fig. 2.13. At higher upset pressure, there is no unmixed region in the welds. The bonding area

Fig. 2.13 Cross-sectional views of the joints showing the effect of upset pressure on flash formation. The flash formation on titanium side varies with increasing upset pressure at P1 = 130 MPa, P3 = 150 MPa and P6 = 180 MPa and deformed to form a ring shape at P3 and P6 [21]

near the interface is found to increase with upset pressure. At higher upset pressure, the bonding area near the interface shows a symmetrical weld metal as seen at P6.

Figure 2.14 shows the micrograph of the weld interface of the sample produced at the lowest upset pressure and the one produced at higher upset pressure. Weld defect was observed at the lowest upset pressure (P1) as a result of inadequate upset pressure and friction time causing the lack of bonding defect shown by the arrow marks. A better bonding was observed at higher upset pressure (P6) in the weld interface without any defects. The better bonding was attributed to the reduction of nickel interlayer thickness as a result of continuous rubbing of abutting surfaces at high upset pressure and higher friction time.

Figure 2.15 shows the thickness of the intermetallic compound layers in the weld interface with an interlayer of 70 lm. The intermetallic compound layer width is more than 70 lm.

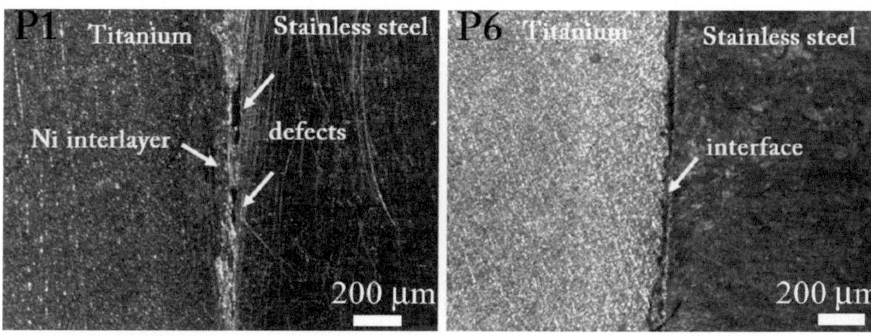

Fig. 2.14 Macrostructures of the joint interface showing the defects formation at low upset pressure P1 = 130 MPa, and a defect-free linear interface at upset pressure P6 = 180 MPa [21]

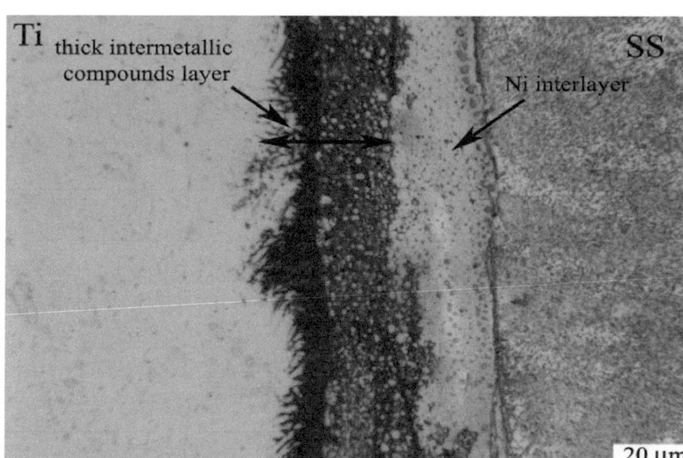

Fig. 2.15 Microstructure of the joint interface at 70 lm thick interlayer showing the presence of thick intermetallic compound formation for the upset pressure P6 welds [22]

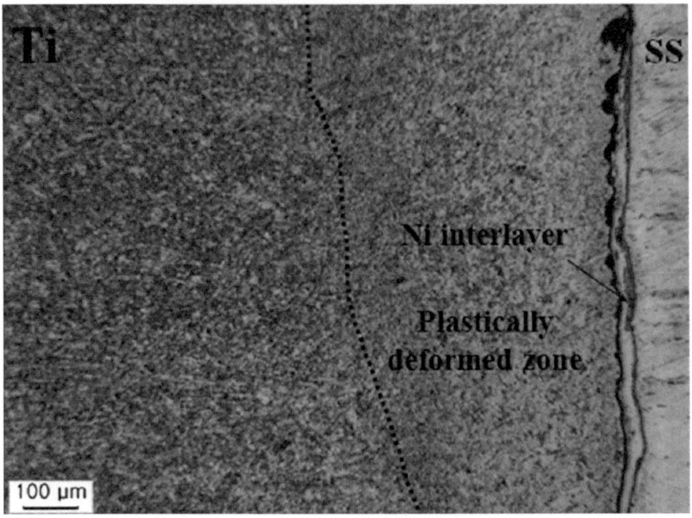

Fig. 2.16 Microstructures of the joint interface showing the plastically deformed region on the titanium side for the upset pressure P6 welds [21]

The micrograph of the plastically deformed region of the titanium side of the sample produced at high upset pressure (P6) is shown in Fig. 2.16. The weld flash from titanium showed a large width of unetched region adjutant to the interface and fine-grained microstructure at the titanium side because of the large deformation of titanium under frictional heat at the interface. This plastic deformation causes the formation of dynamic recrystallization zone with fine equiaxed grains near to the weld interface. The microstructural changes was the result of both the deformation and heat generation at the weld interface which has been dissipated through the substrates causing the temperature gradient that resulted in the formation of different microstructures with dynamic recrystallization that leads to the grain refinement at welds interface.

Lin et al. [22] investigated the influence of a number of solvents on the properties of friction welded PMMA with PVC. Distilled water, methanol, ethanol, acetone and cosolvent (comprising ethanol and distilled water of volume fraction of 10 and 20 Vf%) are the solvent tested. The results of this investigation showed that treatment with those solvents improved the bonding strength of the welded samples of up to 360% more than the samples without solvent treatment. The treatment of samples in the solvents before friction welding allowed for better wetting. Acetone gave the best result followed by ethanol, methanol, distilled water, 20 Vf% cosolvent and 10 Vf% cosolvent. The morphologies of the fracture surface for the untreated sample, the sample treated with distilled water and methanol are shown in Fig. 2.17, Fig. 2.8 and Fig. 2.9, respectively. The fracture surface of the untreated sample of welded PMMA and PVC show three zones as shown in Fig. 2.17: the central zone (Fud), has a "worn surface" morphology in PMMA and PVC samples, the peripheral

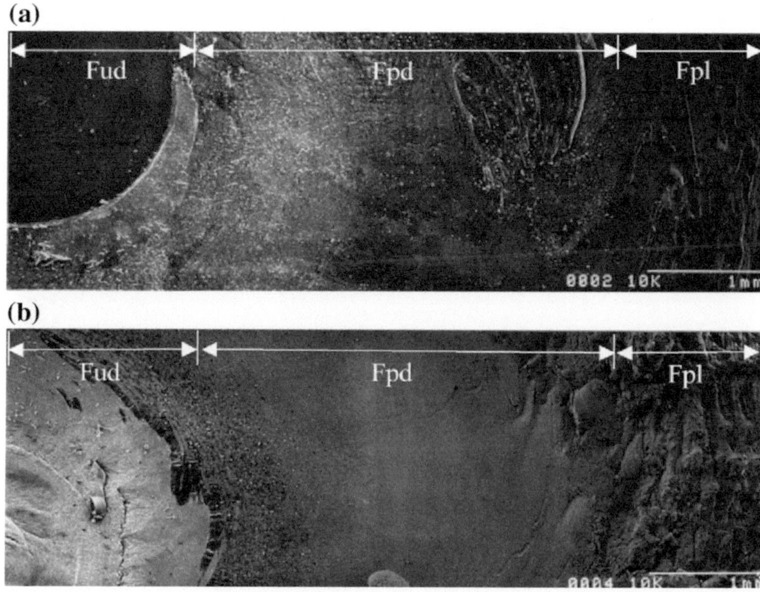

Fig. 2.17 Scanning electron microscope (SEM) morphologies for tensile fractures without solvent treatment: **a** PMMA and **b** PVC [22]

zone (Fpl), has a "rippling" fracture morphology in the PMMA welding part while a "plasticized pile-up" fracture morphology is seen in the PVC welded part, and the middle section (Fpd), that has a partly plasticized "dot" fracture morphology in the both the welded samples. The "dot" fracture morphology can be attributed to the molecular chains of welded PMMA and PVC that penetrate into the matrix of each other and entangle in the overlap zone.

The area of the "dot" fracture morphology seen in samples treated with distilled water as shown in Fig. 2.18 is larger than those without solvent shown in Fig. 2.17. The distilled water is transported to both of the welding parts to activate the partial plasticization of the molecular chains by the friction temperature, which helps to increase the wetting zone for interdiffusion of the molecular chains into one another. The morphology of the fracture surface of the sample treated with methanol showed the largest "dot" fracture morphology for the PMMA as shown in Fig. 2.19 is compared to the ethanol and acetone while ethanol produced the largest effect in PVC followed by methanol and distilled water [22].

Friction welding of the tube-to-tube plate using an external tool (FWTPET) is an attractive friction welding process with applications in aerospace, automobile, marine and railway industries. Dissimilar materials can also be welded using this friction welding process. Senthilkumaran and Muthukumaran [23] produced FWTPET welds with six different tube projections using tool steel as shown in Fig. 2.20. FWTPET has been proved to have the capability of producing high-quality tube-to-tube plate weld joints with improved mechanical and metallurgical properties. The six different

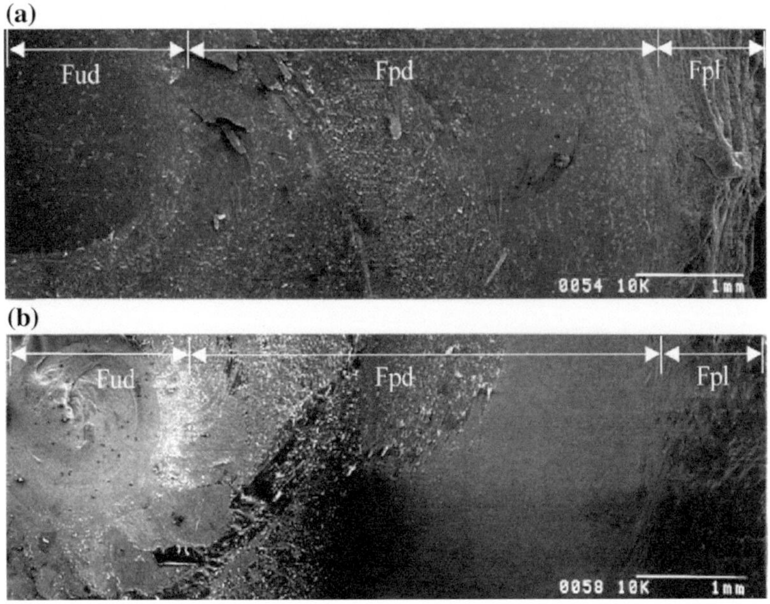

Fig. 2.18 SEM morphologies for tensile fractures treated with distilled water: **a** PMMA and **b** PVC [22]

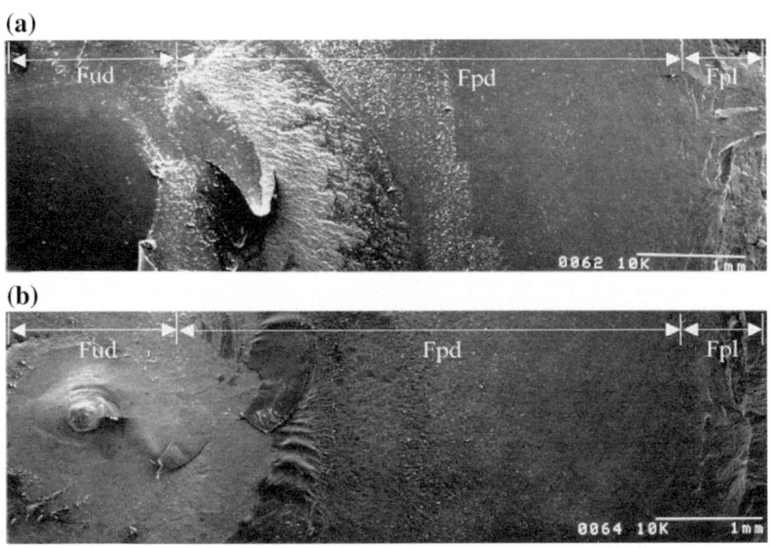

Fig. 2.19 SEM tensile fracture morphologies treated with methanol: **a** PMMA and **b** PVC [22]

Fig. 2.20 Tube-to-tube plate welds (**a** bottom view, **b** top view) [23]

tube projection conditions subjected to FWTPET welding showed that the weld with "1 mm" projection and weld with "2 mm" projection are defect free, while weld produced with "−1 mm" projection, "0 mm" projection, "3 mm" projection and "4 mm" have defects.

Jeong et al. [24] investigated the optimum welding process parameters using finite element simulation and influence of tempering temperature on the mechanical properties of friction welded part of large rotor shaft of diameter 140 mm and a disc of diameter 310 mm using Inertia welding. The process parameters studied include axial force, initial revolution speed and energy, upset pressure and working time. The optimal process parameters determined using the finite element analysis was used to produce the friction-welded joint SANYO special steel and high-strength low-alloy Cr–Mo steel of 140 mm diameter. The sample was quenched in oil at 950 °C for 4 h and tempered in the air at 690–720 °C for 6 h. The results revealed a good agreement between the predicted and the experiment results. Luo et al. [25] studied the properties of current inertia friction welding (CIFW) method that was carried out by hybrid an external additional electronic current in inertia friction welding (IFW) process for fK418 and 42CrMo dissimilar metals. The results showed that the hybrid additional electronic current has a great influence on the properties of the interface of the IFW joints. It also helps to reduce the welding time because of the mixture actions of both friction heat and resistance heat. Also, the widths of the element diffusion zone were found to increase in samples produced with CIFW with an element in 42CrMo diffused into the K418/42CrMo interface. The tensile strength of the CIFW joint was also found to be increased.

Kimura et al. [26] investigated the properties of friction-welded thin-walled pipes of AA6063 aluminium alloy and AISI 304 stainless steel. The pipes with a thickness of 1.5 mm, and the friction-welded joint was made with a friction speed of 27.5 rps and a friction pressure of 30 MPa. The joint, which was made by a continuous drive friction welding machine. In the beginning, the authors observed heavy deformation on the A6063 side during braking. To prevent this from happening, the joint was made by bringing the speed of both workpieces to a stop at the same time after the setting friction time has finished. Using the friction pressure of 30 MPa and a friction time of 0.4 s with 60% weld efficiency, there was no formation of intermetallic compound (IMC) layer (interlayer) at the weld interface. The weld produced at a friction time of 1.6 s has intermetallic. 100% joint efficiency was not achieved because the joint had a softened region in the A6063 side. This region was not reduced from the joint regardless of increasing forge pressure. The joint made at a friction time of 0.4 s with a forge pressure of 135 MPa produces a tensile fracture on the A6063 side. While the joint made with forge pressure of 0 MPa is produced at the adjacent region of the weld interface on the A6063 side. Joint with flash has higher weld efficiency than joint without flash. Sakiyan et al. [27] also studied the influence of friction welding parameters—the forging pressure and welding time on macrostructure and mechanical properties of friction welded valve steel X45CrSi9-3 and Nimonic 80A superalloy. The study revealed that the flash size is increased as the forge pressure and welding time are increased. The sample with the highest strength was produced at a rotation speed of 2000 rpm, forging pressure of 2000 MPa and welding time of 12 s. The result showed that the weld strength is increased by increasing the forging pressure and the welding time. The tensile strength was also found to increase up to a certain value when increasing the forging pressure. The photo of a friction-welded sample is shown in Fig. 2.21 shows severe deformation at the friction interface of the two alloys known as flash. Before a defect-free sample was obtained, the initial trial experiment gave a delamination defect as shown in Fig. 2.22.

A similar study was conducted by Winiczenko [28]. The author investigated the influence of friction welding parameters on the tensile strength and microstructural properties of dissimilar AISI 1020-ASTM A536 joints. A hybrid response surface methodology (RSM) and genetic algorithm (GA)-based technique were used to develop a model for the process. The model was simulated, and optimized to obtain the optimum welding parameters. The investigation revealed that as the friction force and friction time increased, the tensile strength was found to increase. Diffusion of

Fig. 2.21 Macrograph of the X45CrSi9-3/Nimonic 80A [27]

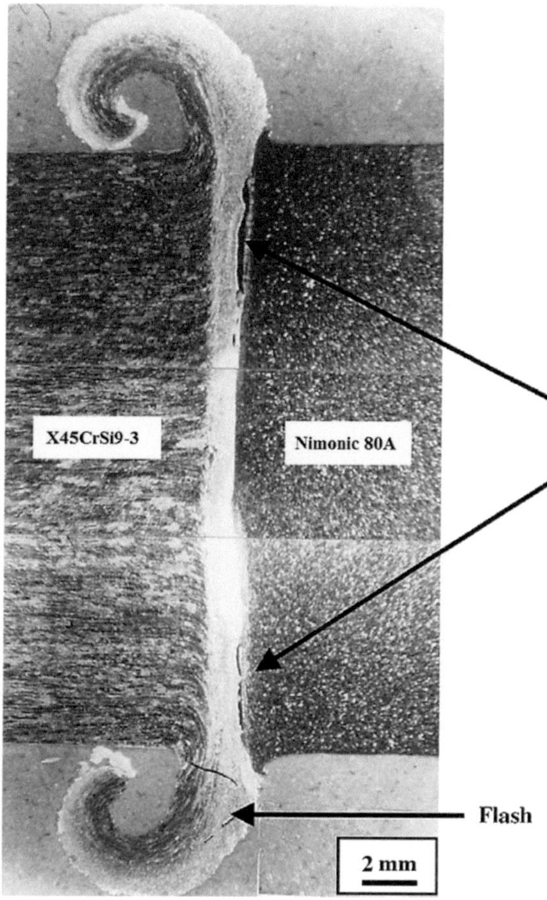

Fig. 2.22 Delamination defects in the joints [27]

carbon atoms from the ductile iron to the steel resulting in the formation of carbon-rich zone and carbon-deficient zones around the interface was observed. In addition, as the upset force was increased, the tensile strength was found to reduce.

Li et al. [29] studied the influence of rotation speed on friction behaviour of friction-welded AA6061-T6 aluminium alloy rods. The results showed that as the rotation speed was increased from 500 to 2100 rpm, the friction time initially increased and then decrease after reaching the maximum value at 900 rpm with the increasing rotation speed. Also, the ratio of heat generation increases from 40 to 70% and then decreases to 19% with the transition point of 900 rpm. This shows that the first stage is crucial in forming the joint below 900 rpm. Łukaszewicz [30] proposed a method of calculation to determine the temperature field during rotary friction welding of metal and formulated an axisymmetric nonlinear boundary value problem of heat conduction of finite length with the assumption that the materials of specimens are thermally sensitive, and the friction coefficient depends on temperature. The solution was obtained using the finite element method with two identical specimens made of AISI 1040 grade steel. The results showed that the lowest temperature in the circular contact region is reached at the centre while the largest temperature was reached on the lateral surface. The thermal sensitivity of AISI 1040 steel was found to be low at the maximum temperature. At a constant friction coefficient, the forging temperature was achieved faster on the entire contact surface. The author concluded that the corresponding thermal stresses and deformations in the weld element can be determined from the found temperature field. A similar studied was conducted by Ji et al. [31] using a 3D coupled thermo-mechanical finite element method to simulate linear friction welding of titanium alloy—Ti6Al4V. The influence of welding parameters—oscillation frequency, oscillation amplitude, friction pressure, and forging pressure—on the material flow behavior was investigated through simulation. The results of numerical simulation showed that the material flow velocity increases with increasing friction time but decreases with increasing forging time. The flow velocity of material in the friction surface increases slowly and then sharply with the increase in friction time at the friction stage. While at the forging stage, the temperature and the flow velocity of material both decrease rapidly. Uzkut et al. [32] conducted a study on the influence of different friction welding parameters applied to two different steels with high alloys on the mechanical and metallurgical properties. The materials studied in this work are X53CrMnNiN219 and X45CrSi93 steel. The results showed that more material loss was obtained with increased friction pressure and forging pressure and at high friction time of 4.7 s. The optimum welding parameters that were obtained in this study are friction time of 3.7 s, friction pressure of 207 MPa and forging pressure of 414 MPa. Faes et al. [33] developed another welding procedure for automatic friction joining of the pipeline using a rotating intermediate ring to generate the heat necessary to produce the weld. The pipes with an outer diameter of 88.9 mm and a wall thickness of 7.6 and 10.0 mm were used for the investigation. The pipe material used was steel API-5L X42 while the welding ring material used was a quenched and tempered steel and a normalized fine-grained steel. The study concluded that the fine-grained normalized steel P355NL1 has a better performance and appropriate for use as welding ring material for welding pipes in API-5L X42.

2.6 Summary

Friction welding is an important solid-state welding process that is used to join metallic material and plastic materials using the process of friction which generates heat that is needed to achieve the welding operation. In this chapter, the process of friction welding, brief background, advantages, limitation and areas of application were presented. The different variants of friction welding processes are highlighted and the processing parameters that govern the friction welding process were described. Some of the research works in the literature are also presented in this chapter.

References

1. A. Orhan, Investigation of the joining characteristics of Al-B4C composites manufactured by friction welding. Sci. Eng. Compos. Mater. **22**(4), 441–448 (2014). https://doi.org/10.1515/secm-2013-0263
2. M. Jan, M. Uday Basheer, A.F.M. Noor, Effect of friction speed on the properties of friction welded Alumina-Mullite Composite to 6061 Aluminum alloy. J. Austr. Ceram. Soc. **52**, 134–142 (2016)
3. M. Srinivasan, C. Loganathan, V. Balasubramanian, Q.B. Nguyen, M. Gupta, R. Narayanasamy, Feasibility of joining AZ31B magnesium metal matrix composite by friction welding. Mater. Des. **32**(3), 1672–1676 (2011)
4. S.K. Sundararajan, S.K. Shanmugam, Multiobjective optimization of friction welding process parameters using grey relational analyses for joining aluminum metal matrix composite. Mater. Sci. (Medžiagotyra) **24**(2) 222–229 (2018). https://doi.org/10.5755/j01.ms.24.2.17725
5. R. Hariharan, G. Prabu, R. Praveen Rackson, V. Pravin Kumar, Rubeshwaren, Study of Hardness in friction welding of aluminium alloy—a review paper, in *2019, International Journal of Management, Technology and Engineering*, pp. 169–179
6. L. D'Alvise, E. Massoni, S.J. Walløe, Finite element modelling of the inertia friction welding process between dissimilar materials. J. Mater. Process. Technol. **125–126**, 387–391 (2002)
7. A. Vairis, M. Frost, High frequency linear friction welding of a titanium alloy. Wear **217**(1), 117–131 (1998)
8. M.F. Zaeh, A. Poehler, Prod. Eng. Res. Devel. **5**, 159 (2011). https://doi.org/10.1007/s11740-010-0284-8
9. R. George Sahaya Nixon, B.S. Mohanty, R. Sathish, Friction surfacing of AISI 316 over mild steel: a characteriation study. Def. Technol. **14**, 306e312 (2018)
10. U. Ashok Kumar, P. Laxminarayana, Friction surfacing process of aluminum alloys, in 5th International & 26th All India Manufacturing Technology, Design and Research Conference (AIMTDR 2014) 12–14 December 2014, IIT Guwahati, Assam, India, pp. 291–292
11. N. Ravi, E. Balaji, S. Rajendra Boopathy, Effect of welding parameters in friction welding of hollow engine valves. J. Eng. Appl. Sci. **11**(12) (June 2016)
12. M. Sahin, Optimizing the parameters for friction welding stainless steel to copper parts. Materiali in tehnologije/Mater. Technol. **50**(1), 109–115 (2016)
13. K.S. Sreenivasan, S. Satish Kumar, Jeeva Katiravan, Genetic algorithm based optimization of friction welding process parameters on AA7075-SiC composite. Eng. Sci. Technol. Int. J. **22**(4), 1136–1148 (2019)
14. N.-E. Titouche, T. Boukharouba, S.A. Amzert, A.J. Hassan, R. Lechelah, S. Ramtani, Direct drive friction welding effect on mechanical and electrochemical characteristics of titanium stabilized austenitic stainless steel (AISI 321) research reactor thick tube. J. Manuf. Process. **41**, 273–283 (2019)
15. G.I. Khidhir, S.A. Baban, Efficiency of dissimilar friction welded 1045 medium carbon steel and 316L austenitic stainlesssteel joints. J. Mater. Res. Technol. **8**(2), 1926–1932 (2019)
16. T. Homma, H. Takano, T. Ozaki, Nanostructural analysis of welded Ti–6Al–4V by linear friction welding applied for blisk assemblies. Materialia **5**, 100174 (2019)
17. B. Cheniti, D. Miroud, R. Badji, P. Hvizdoš, M. Fides, T. Csanádi, B. Belkessa, M. Tata, Microstructure and mechanical behavior of dissimilar AISI 304L/WC-Co cermet rotary friction welds. Mater. Sci. Eng. A (2019). https://doi.org/10.1016/j.msea.2019.04.081

18. D.R. Pissanti, A. Scheidb, L.F. Kanan, G. Dalpiaz, C.E.F. Kwietniewski, Pipeline girth friction welding of the UNS S32205 duplex stainless steel. Mater. Des. **162**, 198–209 (2019)
19. P. Jedrasiak, H.R. Shercliff, A.R. McAndrew, P.A. Colegrove, Thermal modelling of linear friction welding. Mater. Des. **156**, 362–369 (2018)
20. M. Stütza, F. Pixner, J. Wagner, N. Reheis, E. Raiser, H. Kestler, N. Enzingera, Rotary friction welding of molybdenum components. Int. J. Refract. Metals Hard Mater. **73**, 79–84 (2018)
21. M. Cheepu, M. Ashfaq, V. Muthupandi, A new approach for using interlayer and analysis of the friction welding of Titanium to stainless steel. Trans. Indian Inst. Met. **70**(10), 2591–2600 (2017)
22. C.B. Lin, L.-C. Wu, Y.-C. Chou, Effect of solvent and cosolvent on friction welding properties between parts of PMMA with PVC. J. Mater. Sci. **38**, 2563–2570 (2003)
23. S. Senthilkumaran, S. Muthukumaran, Effect of projection on joint properties of friction welding of tube-to-tube plate using an external tool. Int. J. Adv. Manuf. Technol. **75**, 1723–1733 (2014)
24. H.-S. Jeong, J.-R. Cho, J.-S. Oh, E.-N. Kim, S.-G. Choi, M.-Y. Ha, Inertia friction welding process analysis and mechanical properties evaluation of large rotor shaft in marine turbo charger. Int. J. Precis. Eng. Manuf. **11**(1), 83–88
25. J. Luo, L. Li, Y. Dong, X. Xu, A new current hybrid inertia friction welding for nickel-based superalloy K418–alloy steel 42CrMo dissimilar metals. Int. J. Adv. Manuf. Technol. **70**, 1673–1681 (2014)
26. M. Kimura, M. Kusaka, K. Kaizu, K. Nakata, K. Nagatsuka, Friction welding technique and joint properties of thin-walled pipe friction-welded joint between type 6063 aluminum alloy and AISI 304 austenitic stainless steel. Int. J. Adv. Manuf. Technol. (1–4) (2016). https://doi.org/10.1007/s00170-015-7384-8
27. S. Sakiyan, H. Sabet, M. Abbasi, Characterization of mechanical properties in X45CrSi9-3/Nimonic 80A welded by friction welding. Int. J. Steel Struct. **17**(1), 319–324 (2017)
28. R. Winiczenko, Effect of friction welding parameters on the tensile strength and microstructural properties of dissimilar AISI 1020-ASTM A536 joints. Int. J. Adv. Manuf. Technol. **84**, 941–955 (2016)
29. X. Li, J. Li, F. Jin, J. Xiong, F. Zhang, Effect of rotation speed on friction behaviour of rotary friction welding of AA6061-T6 aluminum alloy. Weld. World **62**, 923–930 (2018)
30. A. Łukaszewicz, Nonlinear numerical model of friction heating during rotary friction welding. J. Frict. Wear **39**(6), 476–482 (2018)
31. S. Ji, Y. Wang, J. Liu, X. Meng, J. Tao, T. Zhang, Effects of welding parameters on material flow behaviour during linear friction welding of Ti6Al4V titanium alloy by numerical investigation. Int. J. Adv. Manuf. Technol. **82**, 927–938 (2016)
32. M. Uzkut, B.S. Ünlü, M. Akdağ, Determination of optimum welding parameters in connecting high alloyed X53CrMnNiN219 and X45CrSi93 steels by friction welding. Bull. Mater. Sci. **34**(4), 815–823 (July 2011)
33. K. Faes, A. Dhooge, P. De Baets, P. Afschrift, New friction welding process for pipeline girth welds—welding time optimisation. Int. J. Adv. Manuf. Technol. **43**, 982–992 (2009)

Chapter 3
Friction Stir Welding

Friction stir welding (FSW) is an innovative advanced welding technology that was invented for light metals by The Welding Institute (TWI), England, U.K. in 1991. Since the invention of this joining technology, a considerable weight saving has been attained in industries where the major interest is to save as much weight as possible. Industries such as automobile and aerospace have greatly benefited from this important welding technology. Metals that are difficult to join using the traditional welding techniques are now easily joined using friction stir welding process. Dissimilar materials that were prohibitive to join in the past using the traditional welding process can now be joined using the FSW. This chapter presents the brief history of friction stir welding process, the operating principles of the process, advantages, limitations and areas of application of FSW process are also presented. Extensive research in this field has resulted in the expansion of application area from light metals to other metals. Some of the research works in this field are presented.

3.1 Introduction

Friction stir welding (FSW) is an advanced joining process that uses friction to generate the needed heat required for the joining process. This joining technology does not require filler materials which result in considerable weight saving. Friction stir welding process is continuously gaining attention and the application area is also growing. This welding process is used in a variety of structural applications. Friction stir welding was invented in 1991 by the welding institute and ever since, FSW has gained popularly and it has continuously seen a number of improvements in terms of weld with better fatigue performance when compared with welds produced from traditional welding processes [1, 2]. Dissimilar materials that are wide apart in melting points could not be joined using the fusion welding technology. Friction stir welding makes joining of such dissimilar materials because the melting of the materials is not needed to achieve the joining of the materials. Friction stir welding

E. T. Akinlabi and R. M. Mahamood, *Solid-State Welding: Friction and Friction Stir Welding Processes*, Mechanical Engineering Series,
https://doi.org/10.1007/978-3-030-37015-2_3

process is a solid-state welding process that has revolutionized how metals are joined. FSW has been reported to join aluminium and copper and aluminium alloy and stainless steel [3, 4]. Friction stir welding technology has made it possible to join different materials together effectively for various applications. The main problem faced with the transportation industries such as aerospace and automobile industries is weight saving. All manufacturing processes that will make this a reality is always sought after in these industries. Friction stir welding process enables the joining of light metals and it does not require filler materials that will contribute to the additional weight of the component being joined. The friction stir welding process was invented by Wayne Thomas at the Welding Institute in 1991, in Cambridge, United Kingdom. This process was later developed and eventually patented by the welding institute. Friction stir welding being a solid-state welding process, it produces no molten state materials during the welding process and this results in a considerable energy saving. Also, the heat used in plasticizing the materials is gotten from the friction action in the process. Friction sir welding is a green manufacturing process and an eco-friendly process. The first commercial friction stir welding machines were produced by ESAB1 Welding and Cutting Products at their equipment manufacturing plant in Laxa, Sweden [16]. Since the invention of this process, there has been a significant development in this welding process. Friction stir welding is now used to produce complex assemblies with high quality. FSW now widely used in research and production in many sectors, such as aerospace, shipbuilding, electronic, automotive, housings, coolers, railway, nuclear waste containers and heat exchangers. High level of success has been reported in welding of difficult to weld metals such as aluminium, brass, and copper. The FSW is being in development in terms of research to expand the usefulness of this welding process for high-melting-temperature materials, such as stainless steels and nickel-based alloys through the development of high heat resistance tools that can be used to effectively join these high-melting-temperature materials.

The principle of operation of friction stir welding process is also described in the next section. Advantages, limitations and areas of applications of friction stir welding process are highlighted in this chapter. Friction stir welding process has gained a lot of attention in the research community, some of these research works are presented in this chapter.

3.2 Principle of Friction Stir Welding Process

Friction stir welding is based on the principle of friction and heat generation. A rotating tool comprising of a pin and shoulder is pressed or plunged into the workpiece materials and moved across the faces being joined. The pin is drilled into the materials being joined until the shoulder in the tool makes full contact with the workpiece surface. Once the pin is completely engaged, it is moved and rotated simultaneously in the welding direction as shown in Fig. 3.1. The advancing and rotating motion of the pin and the contacting shoulder of the tool along the seam, results in the formation of an advancing side and a retreating side that heated and softened the material causing

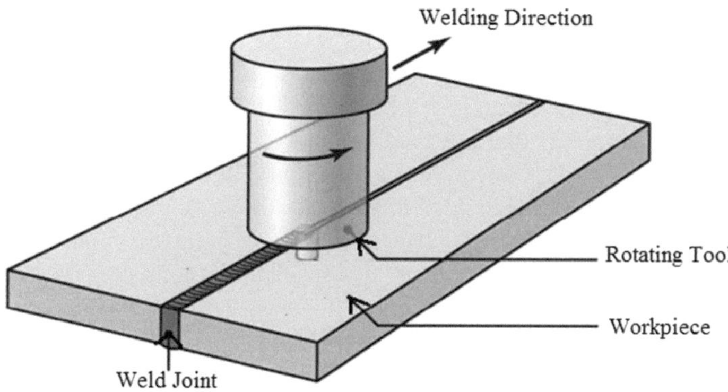

Fig. 3.1 Schematic diagram of friction stir welding process

it to flow around the pin to its backside where the material is consolidated to create form a high-quality, solid-state weld. The rotating tool is used to create the friction and pressure force at the plates. The tool rotation and movement at the plates interface causes friction between the tool and the workpiece which generates heat. The heat generated helps to soften the materials at the interface of the workpiece, which then deforms the interface and diffuses the two workpiece materials into one another by applying a high-pressure force. This creates a strong joint between the interface of the two materials. The welding process is flexible and can be easily automated. The welding process is achieved as follows:

- The plates or sheets to be joined are firmly clamped together with the surfaces to be joined in contact with one another using a heavy-duty setup so that the workpiece does not move or shake during the welding process.
- After proper clamping of the workpieces, the rotating tool pin is then inserted into the workpieces at the interface surfaces between the sheets and plates until the shoulder of the tool touches the workpiece. This action causes plastic deformation to the surface of the materials as a result of the heating action by the frictional force.
- The rotating tool is then moved along the joint line. A strong weld is seen behind the tool as the tool moves across the surface of the workpiece. The tool continues to move unlit the whole weld is produced.
- After the welding process is completed, the tool is separated from the workpiece and finishing operations are performed if required.

The tool used in the friction stir welding process is not consumed during the welding process. It is a wear-resistant material and it is responsible for the frictional heat that is generated during the welding process. This generated heat causes the workpieces to soften but the highest temperature reached is below the melting point of the material. The soften material is transferred from the leading edge of the tool to the trailing edge of the probe that is then mixed and forged together by the action of the tool shoulder and the pin resulting in a solid-phase bond between the two pieces of the workpiece.

3.3 Processing Parameters in Friction Stir Welding Process

One of the important characteristics of friction stir welding process is that there are few process parameters that needs to be controlled unlike in the fusion welding process where many processes need to be controlled. The main processing parameters in friction stir welding process, namely the rotational speed, the travel speed, tool tilt, plunger depth, and the tool design. These processing parameters have an influence on the properties of the weld produced. Each of these parameters is explained as follows.

3.3.1 Tool Rotation and Traverse Speeds

The two important tool speeds that need to be controlled in the friction stir welding are how fast the tool should rotate and how fast the tool should travel across the interface to be joined. These two processing parameters have a great influence on the quality and properties of the weld produced. These processing parameters need to be adequately optimized in friction stir welding process, this is to ensure a successful and efficient welding is achieved. These welding speeds have a direct influence on the resultant heat generation during the welding process. Increasing the tool rotational speed or decreasing the traverse speed will help in the production of a hotter weld. This is a requirement if a successful weld should be made. It is important that the material surrounding the tool should be hot enough to allow an extensive plastic flow of material which will help to minimize the forces acting on the tool. If the materials are not hot enough, voids or other defects will be developed in the weld and it can also cause damage to the tool due to extreme for required by the tool. On the other hand, excessively high heat input is also not good for the final weld properties. This can cause a defect of liquation in low-melting-point materials. Therefore the welding speeds need to be properly controlled in order to produce a weld with the desired properties.

3.3.2 Tool Tilt and Plunge Depth

Tool tilt is another important process parameter in the friction stir welding process that has a great influence on the quality of weld produced. It creates the weld in a slightly lean position or tilt position of between 2 and 4°. The downward forces applied can affect the joint, in order to prevent this condition, a tilt is given to the weld. The plunge depth is the total depth to which the tool gets inserted into the metal sheet. Plunge depth is another important process parameter in friction stir welding process. The plunge depth determines the quality of weld produced. It helps to achieve the desired downward pressure needed in friction stir welding process.

It also ensures that the tool fully penetrates the weld so that the machine may not deflect from its position. An excessive plunge depth is also not good for the weld because it leaves the pin rubbing marks on the weld surface. The plunge depth needs to be set adequately to ensure that the necessary downward pressure is achieved and also to ensure that the tool fully penetrates the weld. If the correct depth is not set for the friction stir welding process, the high pressure required in the welding process may cause the machine to deflect and affect the weld quality.

3.3.3 Tool Design

Tool design is another very important process parameter that greatly affects the quality of the weld produced in friction stir welding process. There are different types of tools design that are used in this welding process and each has an influence on the quality of weld achievable. The tool material should be sufficiently strong with high wear resistance property in order to produce a weld with a good surface finish. The tool should also conduct less heat in order to reduce heat losses.

3.4 Advantages, Limitations and Areas of Application of Friction Stir Welding Process

Stir welding process is an economical joining process because it is not an energy-intensive welding process that uses about 20% heat input when compared to traditional gas metal arc welding processes. This welding process requires minimum surface preparation. It does not produce fumes or toxic gases that could have an adverse effect on the health of the operator, and other personnel. The major advantage is in the welding of dissimilar materials and alloys such as copper and aluminium. Aluminium are now welded with excellent properties using the friction stir welding process. Other non-weldable materials can now be welded using FSW process. This welding process has demonstrated a number of advantages over the competing traditional welding processes and some of these advantages are stated as follows:

- Being a solid-state process, there is no problems of hot cracking and porosity.
- No filler material or shielding gas is required for this welding process.
- There is no production of fumes and unwanted metal spatter of the molten state which is commonplace in the fusion welding process. There is no ultraviolet radiation, in this welding process, hence it is environmental friendly. Friction stir welding can also be regarded as a "green" technology because it does not produce any health hazardous materials which may lead to the destruction of the environment or people.
- The welding process can be easily automated which reduces the need for skilled welders.

- Weld can be produced in any position.
- The mechanical properties of the metal are not degraded after friction stir welding process and in some cases, the properties are actually improved.
- It can be used to weld 'unweldable' aluminium alloys.
- It can be applied to different types of joints such as lap joints, butt joints, T joints, fillet joints, etc.
- This welding process produces fine grain sizes resulting in the high-quality weld.

The limitations of friction stir welding process include the following:

- The exit hole that is left after tool withdrawal after the welding process is complete.
- It requires a significant downforce and traversing forces, therefore the workpiece needs to be clamped more substantial than for arc welding.
- Friction stir welding cannot be used where material deposition is desired.
- Initial investment of the FSW machine is too high.

The high level of success recorded in friction stir welding process within the shortest space of time is responsible for the various application areas of the welding process. These application areas keep growing day in, day out. Friction stir welding process finds its application ranging from micro-welding of electronic components to the heavy-duty fabrication of automotive and aerospace components. This welding process was originally limited to low-melting-temperature materials, with development of new high strength tool materials such as polycrystalline cubic boron nitride (PCBN), tungsten rhenium, and ceramics, higher temperature materials such as steels and other high-strength materials are now being welded using this welding process. As material development continues, different tool materials will continue to be developed to further expand the application areas of this technology. Some of the areas of application of FSW process are as follows:

- **Marine Industries—**The marine industries have adopted the process friction stir welding for commercial applications to weld big parts of the ship. It is used in the building of ship parts such as panels for decks, bulkheads, floors and superstructures. Helicopter landing platforms and submarine structures.
- **Space Shuttle—**The space shuttle's gigantic external tank is now built using friction stir welding, and with the successful development of new alloy—aluminium–lithium Al–Li 2195, that is lighter in weight, the external tank is now built with the new alloy using friction stir welding process to build the shuttle's super lightweight tanks. This is because friction stir welding produces stronger welds that are easier to produce.
- **Nuclear Plant**—As nuclear plants around the world grow older, there are increasing incidences of stress corrosion cracking (SCC) problems. This corrosion cracking problem is caused as a result of pressure in the reactor pressure vessel parts and the water at the reactor nozzle. Friction stir welding is employed to repair the parts affected by the SCC. Frictions stir welding gives a high integrity smooth repair when applied.

- **Aerospace Applications**—Friction stir welding process can be used in the repair of ageing aircraft, fabrication of new structures, and tooling for assembled structures of aerospace parts. Friction stir welding can offer significant advantages when compared to riveting in terms of weight savings and lower manufacturing cost. Friction stir welding process is mostly used to join bigger parts like wings, fuselages, empennages, cryogenic fuel tanks for space vehicles, aviation fuel tanks and other bigger structures in the aerospace industries. Friction stir welding process can also be used to increase the size of the commercially available sheets by joining them together before carrying out other manufacturing processes.
- **Automotive Industries**—It is used in automotive industries to weld many parts together such as to weld wheel rims, chassis, fuel tanks and other structural work to reduce the weight that helps to increase fuel economy. Other parts that can be joined including engine and chassis cradles, truck bodies, tail lifts for lorries, mobile cranes, armour plate vehicles, fuel tankers and caravans.
- **Others**—Friction stir welding is used in the railway industry for building of train cabins for joining the panels together. It is also used in chemical industries for joining heat exchanger, pipeline, etc. Friction stir welding is used in the electronic industries for joining aluminium to copper, connectors and other electronic equipment, the bus bar, etc. Friction stir welding is also used in the fabrication industries in the fabrication of metals.

3.5 Friction Stir Spot Welding

Friction stir spot welding is another variant of friction stir welding process. Friction stir spot welding process is a solid-state single-spot joining process made between adjacent materials at overlap configuration as shown in Fig. 3.2 [5]. The four important steps in the friction stir spot welding process has four basic steps that are shown in Fig. 3.3. The plates to be joined in clamped together as shown in Fig. 3.3, then the rotating tool is plunged into the surface of the top sheet. The tool continues to rotate even after the tool has plunged completely into the plates while pressure is also applied for a length of time. This causes materials around the tool pin to be stirred and mixed together making the lapped plates to be metallurgically bonded together after which the tool is withdrawn from the sheets.

Friction stir spot welding finds its application in transport industries such as automotive, rail and aerospace industries because of the high strength of weld produced making the welding process to be used on parts that are subjected to high load and the weld is able to withstand high loads 7. In the electrical industry, friction stir spot welding can be used to join aluminum and copper together. Friction stir spot welding is used to also join a variety of materials [6–14].

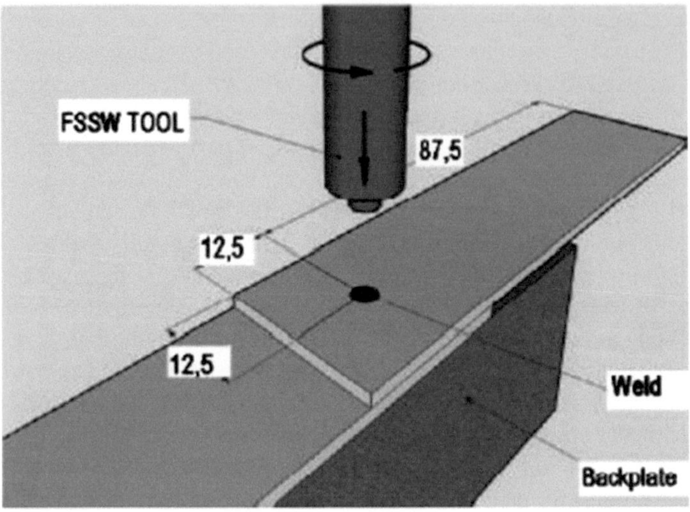

Fig. 3.2 Schematic illustration of FSSW [5]

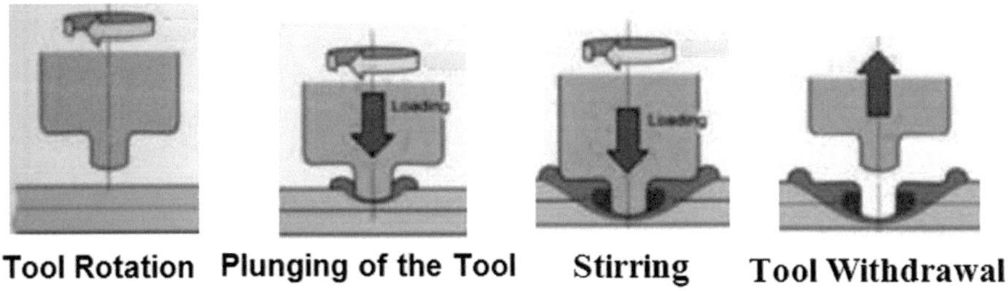

Fig. 3.3 Steps in friction stir spot welding process

3.6 Research Advancement in Friction Stir Welding Process

Friction stir welding has gained a lot of attention in the research community since the advent of this important manufacturing technology. The technology is continuously evolving and has seen a number of improvements as a result of these research activities. Some of the research works in this field are presented in this section. The fatigue behaviour of friction stir welded single-sided aluminium butt joints was investigated under bending and tension load using small-scale specimen testing and fracture mechanics analysis by Ranjan et al. [15]. Four-point bending and tension loading are used to study the fatigue behaviour of the friction stir welded joints. The study also took into consideration the effect of angular misalignment on the fatigue behaviour of the weld. The angular misalignment can occur as a result of differential

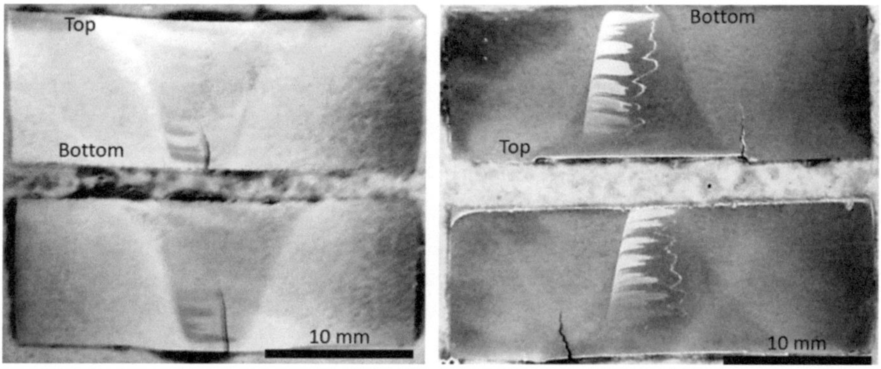

Fig. 3.4 Cross section of welds after bending tests for specimens subjected to tension at bottom of the weld (left), specimens subjected to tension at top of the weld (right) [15]

cooling or inadequate clamping during welding. This misalignment can have a significant influence on the fatigue performance of the welded joint due to the secondary stresses that such misalignment can cause. A strain-based fracture mechanic model was employed in the study. The result of this investigation revealed that the fatigue performance of the welded joints showed a better fatigue performance than a typical arc-welded joints.

The micrograph of the cross sections of the friction stir welded specimens that were tested in two different orientations is shown in Fig. 3.4. The image shown on the left-hand side of Fig. 3.4 is for the specimens that were tested in bending at the top and tension at the bottom of the weld. The crack was seen to have been initiated at the joint zone. The specimen subjected to tension at the top and bending at the bottom is seen to have failed at the interface of the stir zone and the parent material. This was attributed to the slight change in thickness that caused small stress concentration at the crack initiation site.

One of the advantages of friction stir welding process is the ability to join dissimilar materials. Shadeghian et al. [16] studied the flow behaviour of dissimilar materials during friction stir welding process with the aim to be able to predict such material flow of dissimilar materials during friction stir welding process. Thermal and computational fluid dynamics (CFD) simulations were used to determine the temperature distribution and material flow velocity during the welding process. The authors used level set method for morphological simulation of the weldment to predict the weld morphology based on the CFD results obtained. The weld morphologies were studied through simulation using different tool rotational speeds, offsets position and height levels of the specimen of steel and aluminium welded together. The simulation result was validated experimentally. The simulation results and experimental results are in good agreement with one another. The simulation results showed that by increasing the rotational speed and offset through the steel side could generate more steel particles and the steel particles were more in the top height level of the weld section than in the middle and bottom levels. The main cause of defects in the weld zone was seen to be caused by the steel particles detached into the aluminium matrix as shown in Fig. 3.5. It is seen that the cause of defect during friction stir welding of aluminium and steel is due to the spreading of the steel particles into the matrix of the aluminium which is responsible for the microscopic cracks that are formed in the interfaces of the two materials because the two materials have significantly different properties that consequently causes the weld defects. The study showed that this defect can be reduced by having pin offset through aluminium to reduce the steel particles that is distributed in the stir zone. Also, different height levels of the weld section lead to different flow behaviours. The upper part level of the weld has more steel particles than the middle level of the weld, while the bottom level of the welding has less material flow and mixing of dissimilar metals.

Cunha et al. [17] studied the effect of welding speed on the mechanical and microstructural properties of friction stir welded GL E36 shipbuilding steel. The welding speed was varied between 1 and 3 mm/s and the tool rotational speed was kept at a constant value of 500 rpm throughout the experiments.

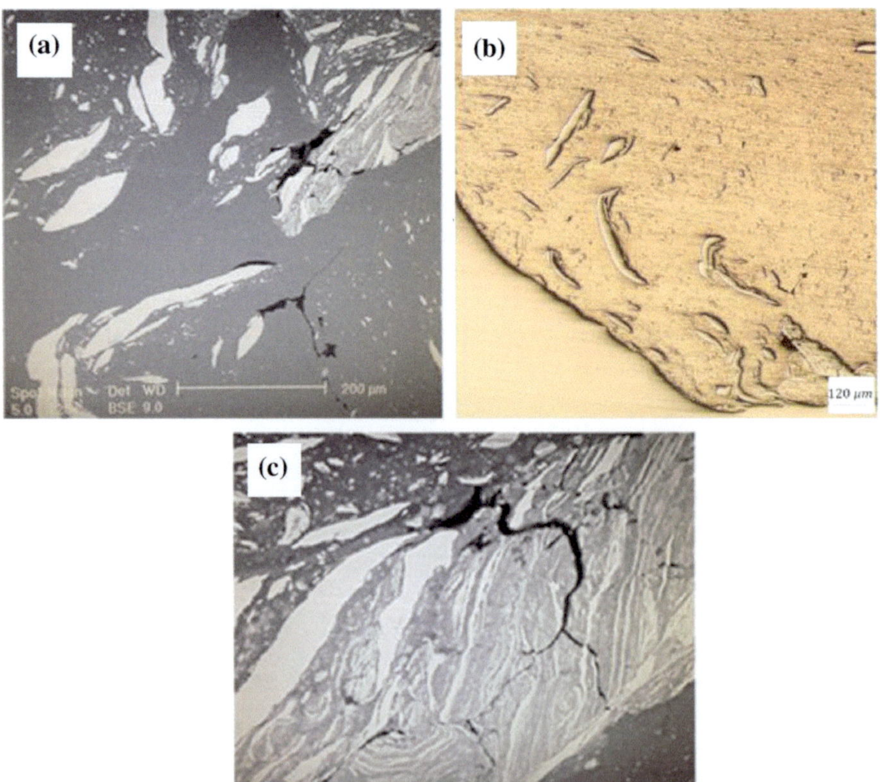

Fig. 3.5 **a** SEM micrograph of steel particles in aluminium matrix, **b** Optical microscopy of steel particles in aluminium matrix, **c** crack propagation beside steel particle [16]

The result of this investigation showed a very good quality weld that indicates a stability of the process parameters used in this study. The macrographs of the weld showed different material flow patterns depending on the heat inputs based on the welding speed used. Large grains are shown within the stirred zone and all the tensile samples did not break at the weld joint but broke at the base material showing that the weld joints have higher strength. When a high welding speed of 3 mm/s lack of penetration was observed which could be improved by increasing the axial force. Figure 3.6 showed the microstructures of the centre of the stirred zones of the weld joints. Complex mixed of microstructures consisting of ferrite (F), martensite (M) and bainite (B) as indicated on the micrograph in Fig. 3.6. A low welding speed results in higher welding temperatures which produced a larger austenite grain. The microstructure was found to consist of ferrite, martensite and bainite showing different levels of grain refinement. The photographs of the bent samples are shown in Fig. 3.7. The sample processed at 1 and 2 mm/s welding speeds were able to be completely bent due to absence of defects within the joint. The sample processed at 3 mm/s welding speed shows defect at the joint root of the bent specimen. The failure in root bending will lead to poor fatigue performance. This shows that when higher welding speeds are used, the axial load should be increased to ensure proper material stirring and bonding.

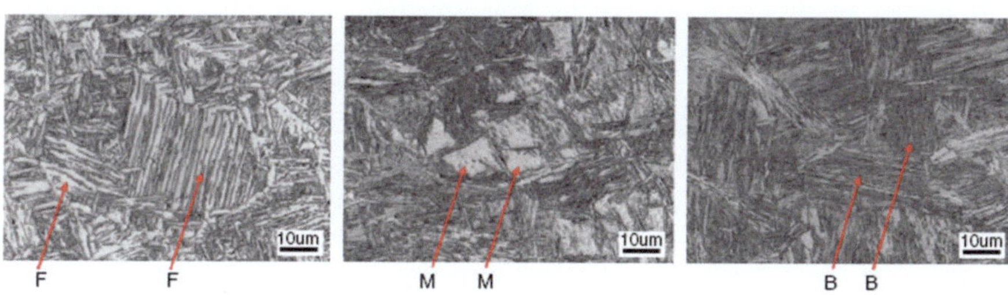

Fig. 3.6 Complex mixed microstructures of the mid-thickness from the centre of the stirred zone, ferrite highlighted as "F", martensite as "M" and bainite as "B" [17]

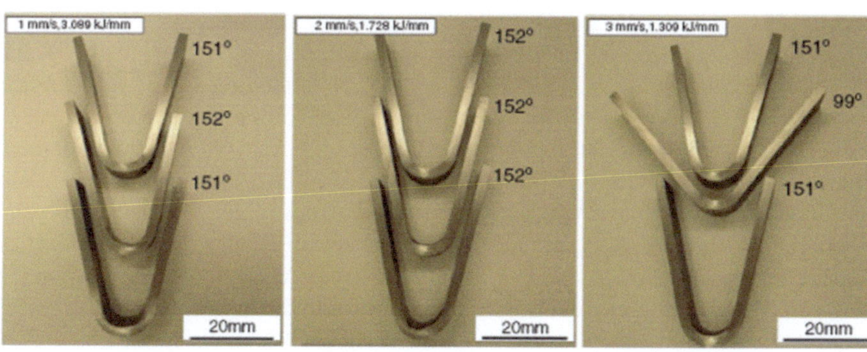

Fig. 3.7 Bent samples, heat input and the measured angle for each welding speed condition [17]

Chen et al. [18] studied the influence tool rotational speed on grain size, grain orientation and precipitates and their effect on mechanical properties of friction stir welded 5A06 aluminium alloy. The result of this study showed that recrystallization occurred in the weld zones, which is replaced by the shear texture components when the rotational speeds are increased. The development of this shear texture, with the precipitates dissolution, and the dislocations pinned by precipitates may cause the coarsening of the grains in the weld zones when higher tool rotational speeds are used. The recrystallization that occurred at low welding speed s during the welding process results in the formation of fine equiaxed grains in the weld zones. The equiaxed grains are replaced by the shear texture in the weld zones at higher tool rotational speeds of 700 and 800 rpm. At higher rotation speed of 700 rpm, the thermo-mechanically affected zone (TMAZ) is considerably smaller than those observed a rotation speed of 400 rpm due to the disappearance of the big round precipitates of Al_3Mg_2 and the precipitation of small round plate-like precipitates Al_6FeMn at high rotation speed which was dissolved in the nugget zone (NZ). The precipitate of Al_6FeMn was seen to dissolve partially at a higher rotation speed of 800 rpm, but the rod-shaped precipitates of Mg_2Si was seen to increase and coarsened when the tool rotation speed was increased from 400 to 700 and 800 rpm as shown in Figs. 3.8 and 3.9. The grain coarsening at higher rotation speeds could be attributed to the shear texture, precipitates dissolution and dislocations pinned by precipitates.

A similar study was conducted by Mahmoudiniya et al. [19], by varying the tool rational speed between 600 and 1000 rpm. 2 mm = thick ferrite–martensite DP700 steel sheets was friction stir welded at the varying rotational speeds and the influence of the rotational speeds on the microstructure and mechanical properties of the welds were investigated. The result revealed that by increasing rotational speed results in grain coarsening in the stir zone. Also, increasing the rotational speed causes an increase in the material softening phenomenon in the sub-critical heat-affected zone. Tungsten carbide (WC) particles were seen in the stir zone, due to the tool wear, and the formation of soft ferrite band that reduces the tensile properties at rotational speed of 600 rpm. The softening at heat-affected zone was responsible for the reduction of the strength and ductility in at higher rotational speed of 800–1000 rpm. Figure 3.10 shows the microstructure in the inter-critical-heat-affected zone (IC-HAZ). The microstructure consists of ferrite and martensite phases. This shows that the temperature in this region has reached a value between Ac1 and Ac3 during the welding process. The higher carbon content in the steel causes the formation of martensite during the cooling cycle. The volume fraction of martensite increases with an increase in rotational speed. Figure 3.11 shows the microstructure of the centre of stir zone of the different rotational speeds. The microstructures consist of the primary ferrite (PF), the Widmanstätten ferrite (ferrite side plates (FS)), acicular ferrite (AF) and bainite. The presence of primary ferrite, as well as bainite, shows that the peak temperature in SZ was well above Ac3 temperature during the friction stir welding process which causes the dynamic recrystallization of the austenite phase before cooling.

Batrand et al. [20] studied the influence of tool rotational speed and advancing speed on the properties of friction stir welded aluminium alloy using the range of

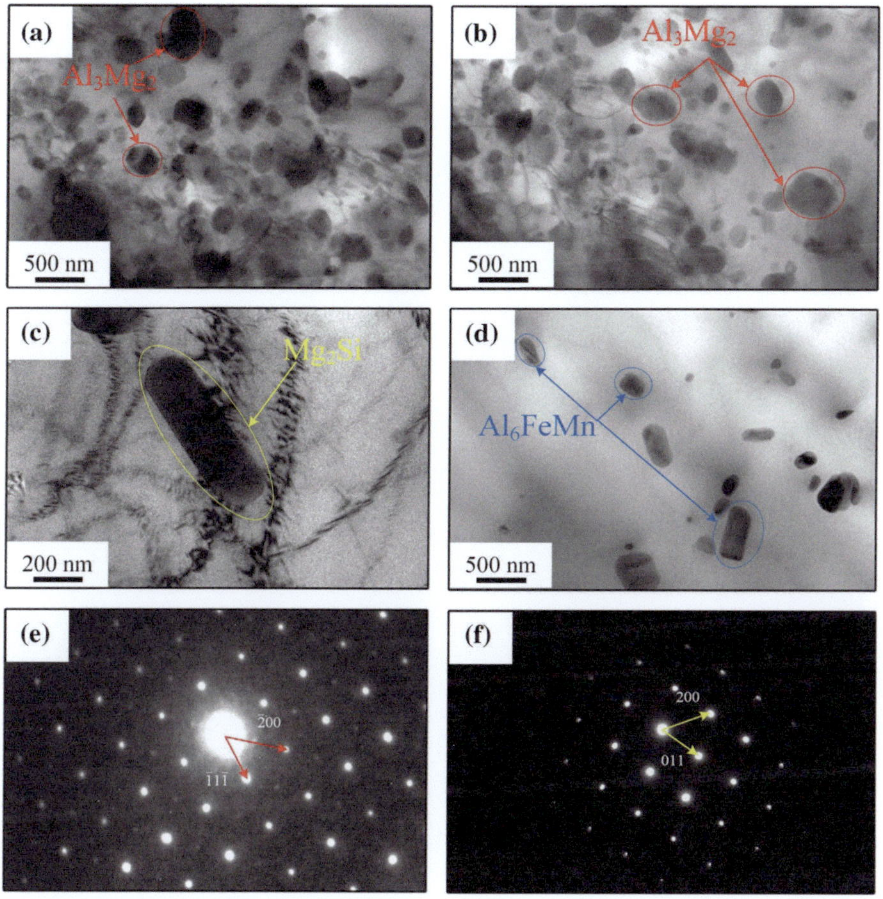

Fig. 3.8 The TEM images in the NZ at a rotation speed of 400 rpm displaying the big round precipitates Al_3Mg_2 and the rod-shaped precipitates Mg_2Si with corresponding diffraction patterns in **e** and **f**, respectively [18]

600–1000 rev. min^{-1} and 250–550 mm per min as rotation speed and advance speed, respectively. The study revealed that the higher the advancing speed, the smaller the width of the minimum hardness region (MHR) is and the closer it is to the nugget region. The lower the advancing speed, the higher the exposure time to the critical temperature ranges. The exposure time to this critical temperature range was also found to could play a significant role in controlling the hardness and the width of the MHR. The fractures in the MHR are similar for all joints and are characterized by dimples and transgranular fracture as shown in Fig. 3.12. At higher advancing and rotation speed condition, the specimens failed in the nugget zone was most likely to be as a result of a lack of cohesion at the interface between the two materials in this area due to lack of material mixing in this region causing lack of metallurgical bonding. This was suggested can be avoided by changing the tool geometry and increasing

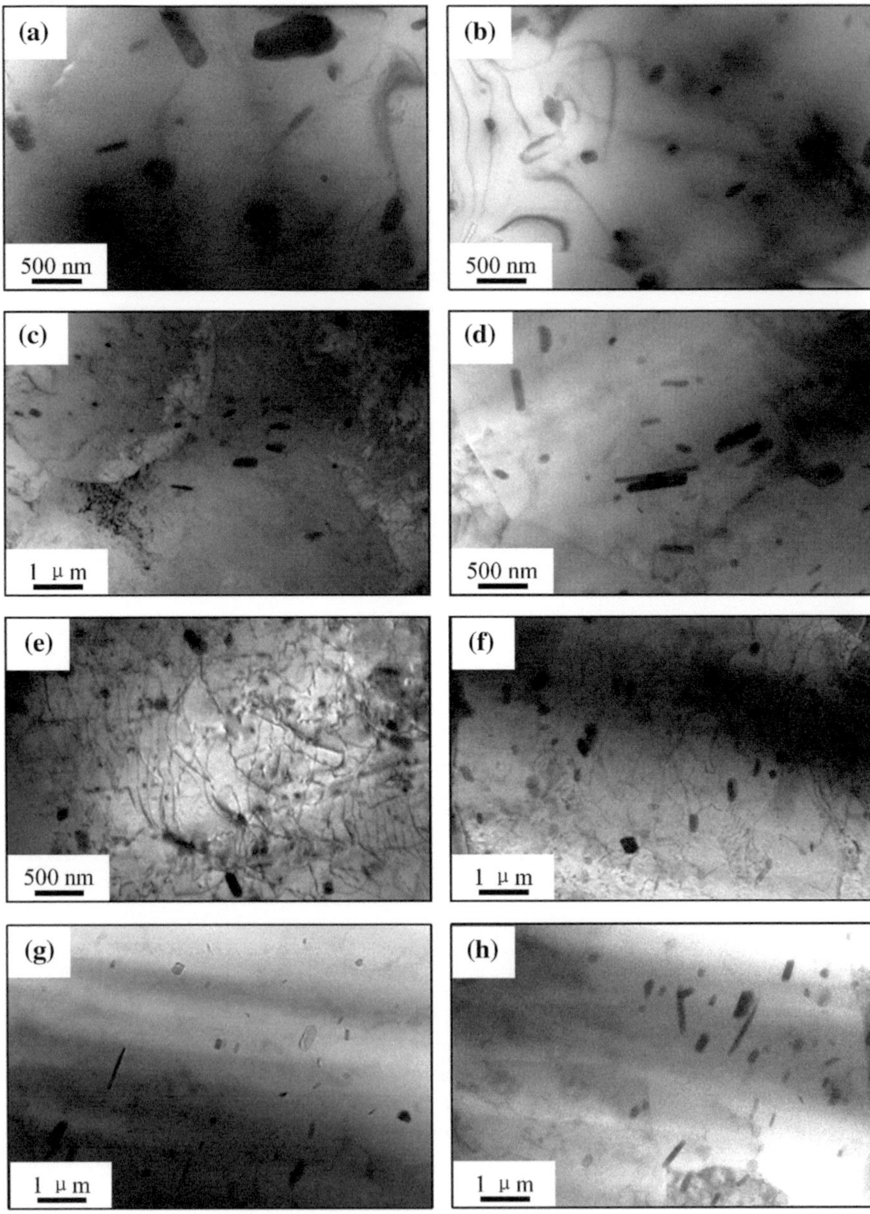

Fig. 3.9 The TEM images in the TMAZs at rotation speeds of 700 rpm and 800 rpm, respectively. **a**, **b** are the TEM images at the RS TMAZ at 700 rpm, **c**–**e** are the TEM images at the AS TMAZ at 700 rpm, **f**–**h** are the TEM images at the AS TMAZ at 800 rpm [18]

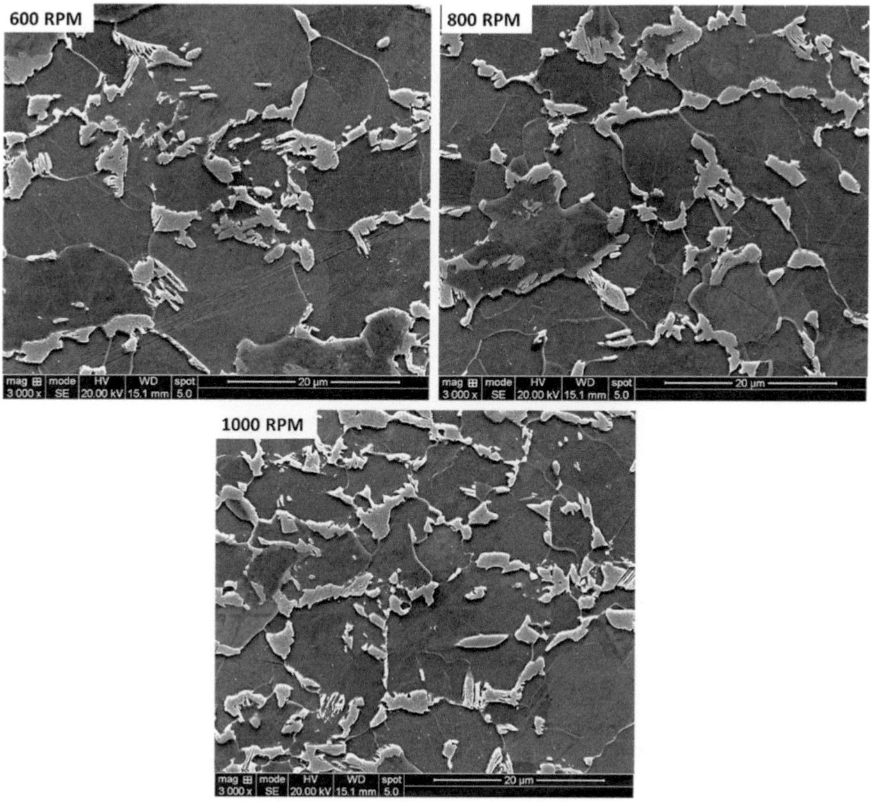

Fig. 3.10 SEM images for IC-HAZ at various rotational speeds [19]

the length of the pin for higher welding speed conditions. The poor performance in tensile strength and elongation of friction welded joint at higher rotational speed condition was attributed to insufficient mixing of the materials at these root of the joint because of the small pin size that result in smaller width of the of the minimum hardness region (MHR) and closer nugget region [20]. The MHR is the locations of significant necking during fracture.

Tang et al. [21] investigated the microstructure and properties of friction stir welded China low activation martensitic (CLAM) steel to 316 L steel. The authors achieved defect-free welded joints by placing 316 L stainless steel at the retreating side (RS) (in case 1) and by placing 316 L in the advancing side (AS) (in case 2). The investigation revealed that quenched martensite was seen in the microstructure of the weld on the CLAM side with coarse grains in the HAZ that result in softening of the zone that is more pronounced in case 2. Uniform austenite microstructures was observed in the microstructure of the 316 L side of the weld. The maximum hardness zone is seen at the nugget zone on the CLAM side. The tensile strength of the welded joints is high and reached the tensile strength of the 316 L base metal. The tensile

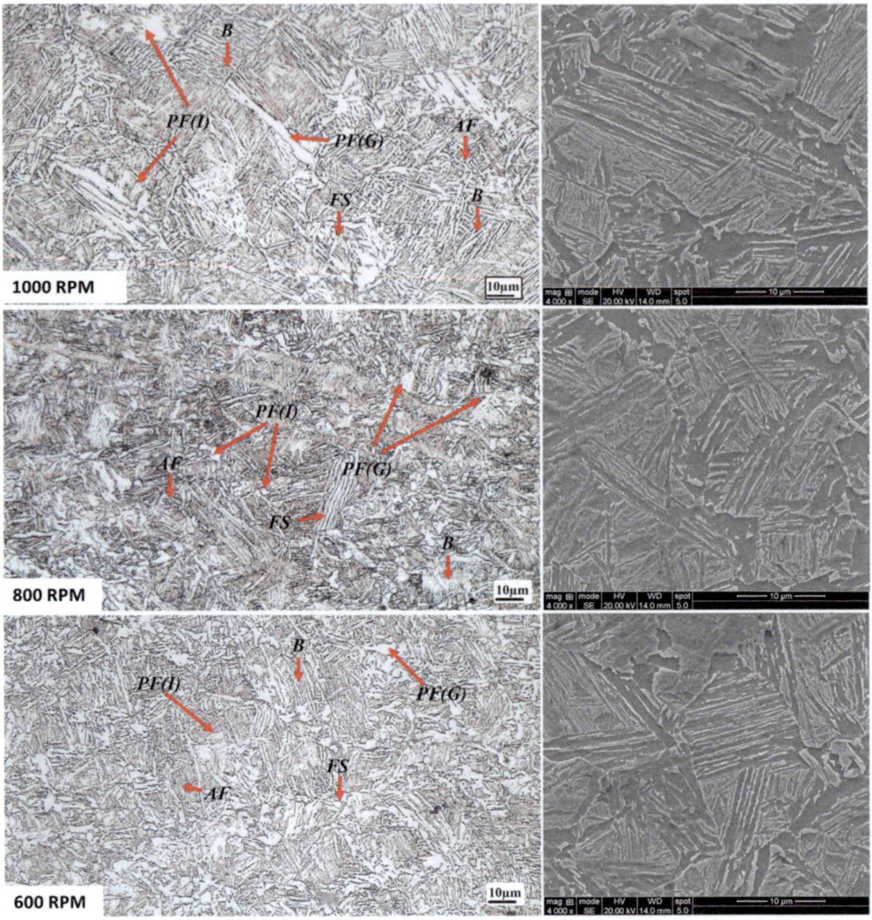

Fig. 3.11 Optical micrographs and FE-SEM images of SZ at various rotational speeds (PF(G): grain boundary ferrite, PF(I)): intragranular ferrite, (FS): Widmanstätten ferrite, AF: acicular ferrite and B: (bainite) [19]

properties of the welded joint obtained in case 1 are better than in case 2. The bonding strength of interfaces between CLAM and 316 L steels in the cases considered were excellent. The microstructure of the HAZ, RS, AS and stir zone (SZ) is shown in Fig. 3.13. The precipitate of carbides in the intragranular regions of the inner HAZ dissolve at the grain boundaries as seen in Fig. 3.13a and b. Also, quenched martensite forms at the grain boundaries during the subsequent rapid cooling process is shown in Fig. 3.13c and d.

Gao et al. [22] developed a finite element model based on solid mechanics to study the material flow in friction stir welding process. The simulation results showed that the flow pattern of the tracer particles around the pin has spiral movement. The material on the upper surface of the weld has a spiral downward movement that is

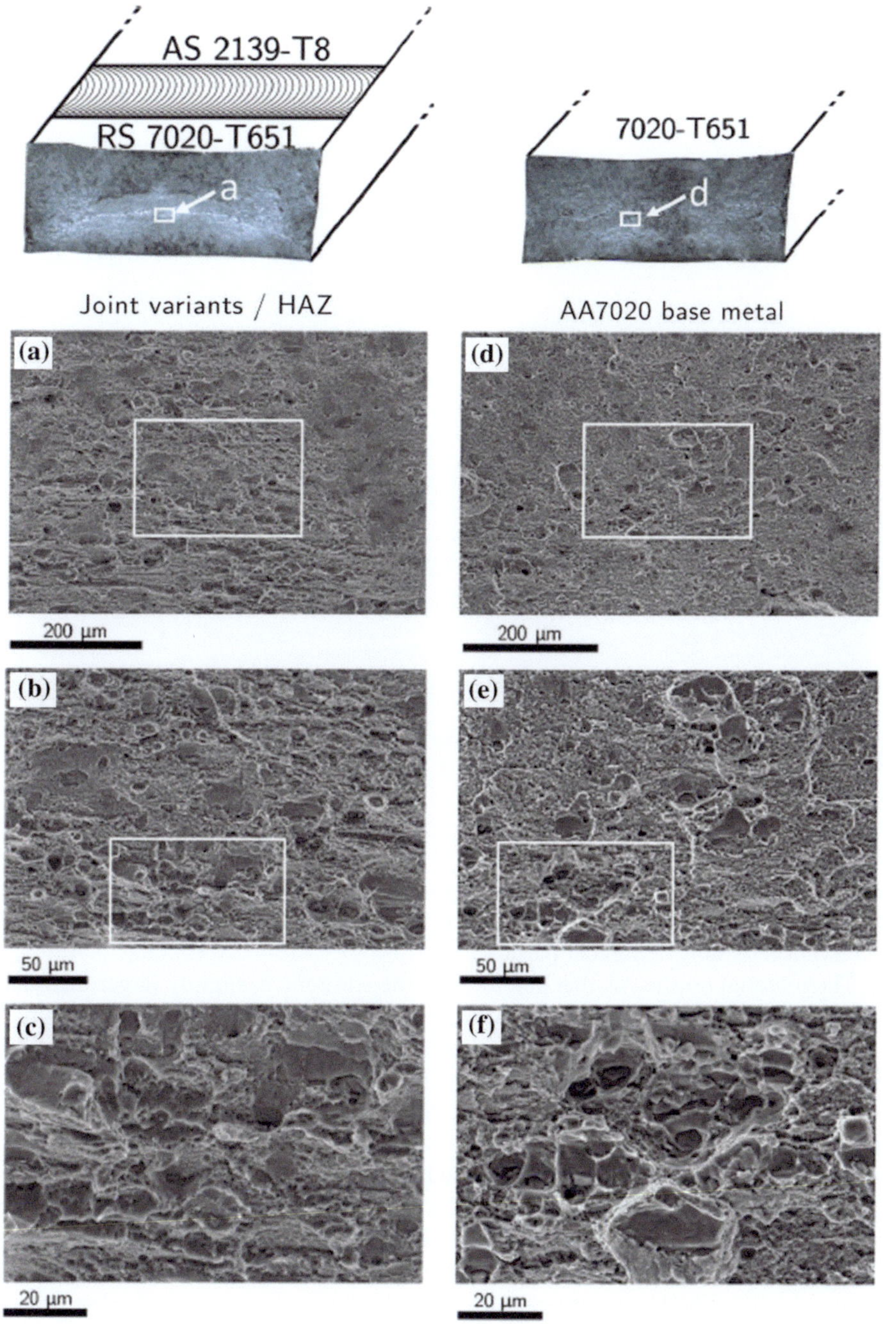

Fig. 3.12 SEM fracture surface features at several magnifications. **a–c** welded joint (condition BW03): mixed-mode fracture composed of dimples initiating from small intermetallic particles and transgranular fracture from elongated grains, **d–f** AA7020 base metal—dimples close to each other and only elongated grain features [20]

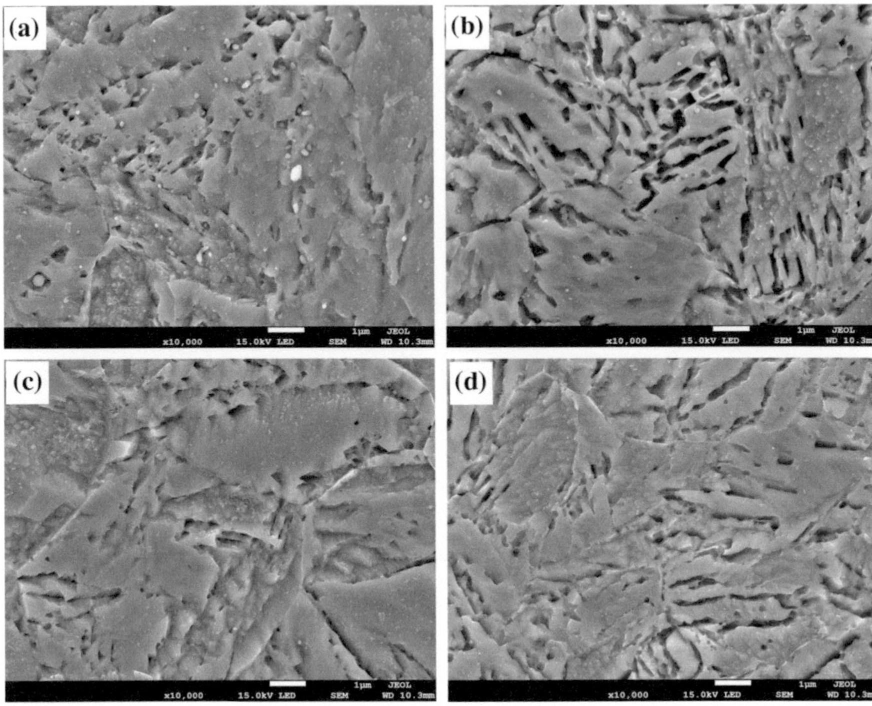

Fig. 3.13 SEM images of **a**, **b** inner HAZ; **c**, **d** SZ; **a**, **c** 316 L at the RS; **b**, **d** 316 L at the AS [21]

affected by the shoulder while the lower material has the spiral upward movement that is affected by the pin. The higher velocity of material flow is observed at the sides of the stirring pin while the velocity of material flow is lower at the bottom of the stirring pin. This result was found to be in good agreement with the experimental observation by tungsten tracer particle.

The material on the AS that entered into the stir zone is slightly larger than that entering the stir zone from the RS at the same distance from the centre of the pin.

Huanh et al. [23] studied the joint formation mechanism of high depth-to-width ratio in friction stir welding of aluminium profiles using a numerical model of plastic flow and experimental approaches. The authors proposed a fluid–solid interaction algorithm to establish the coupling model, and the material to be welded was treated as non-Newtonian fluid. The study showed that the thread structure and the milling facets greatly increased the strain rate under the extremely low heat input that helped to prevent the welding defects. The condition of the peak temperature of 648 K and the strain rate of 151 s^{-1} was attributed to the lowest coarsening degree of the precipitate. The tensile strength of the weld obtained is up to 265 MPa, which is equivalent to 86% of the base material. The comparison of defect distributions between experiments and simulations at different welding speed is shown in Fig. 3.14. The voids with an average diameter of 19 μm were seen distributed at the root of the

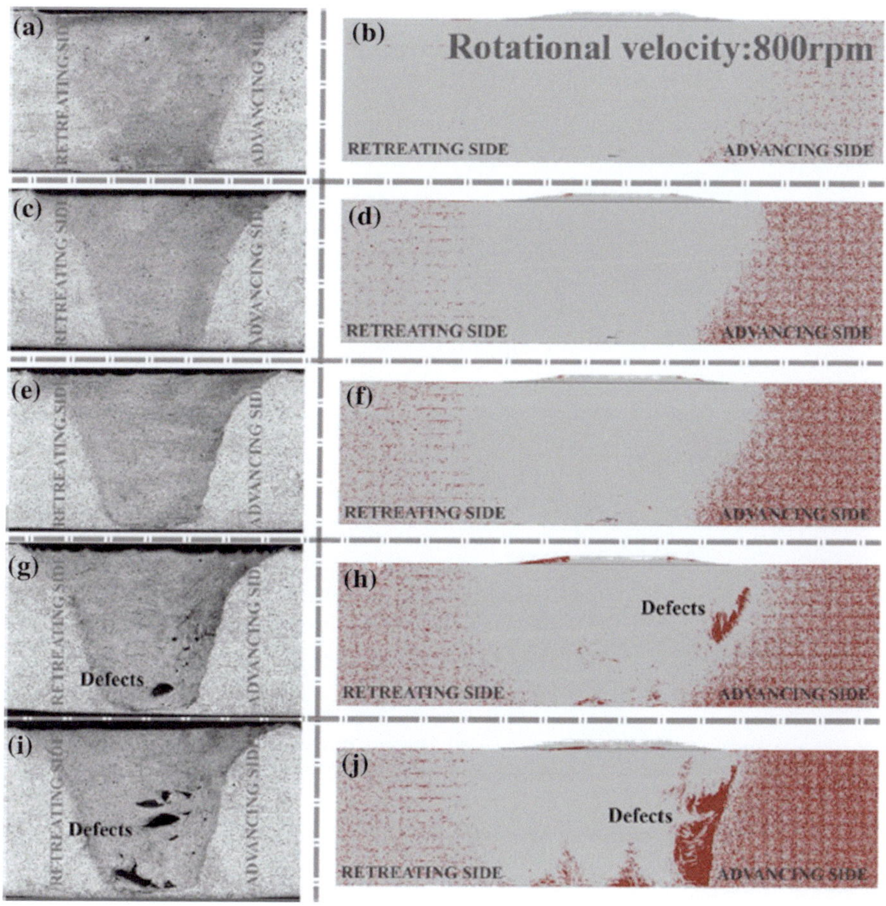

Fig. 3.14 Comparisons of defect distributions between experiments and simulations: **a** and **b** 30 mm/min; **c** and **d** 100 mm/min; **e** and **f** 300 mm/min; **g** and **h** 400 mm/min; **i** and **j** 600 mm/min [23]

NZ in the sample produced at 300 mm/min. Many of such voids are also seen to appear and penetrate each other at the welding speeds of 400 mm/min and 600 mm/min but there was no welding defect seen in the image in Fig. 3.14e. The micrographs of the fracture surfaces at different welding speeds is shown in Fig. 3.15, which shows that the nucleation of the voids in that sample occurs during the tensile testing process and not from the welding process as shown in Fig. 3.15b. The amelioration through the material flow model developed helped to inhibit the welding defects and helped to optimize the process parameter that provides references to extracting process–structure–property linkages for friction stir welding process.

Sinhmar and Dwivedi [24] also investigated the effect of welding speed on weld thermal cycle, the maximum temperatures at low and high scanning speeds and

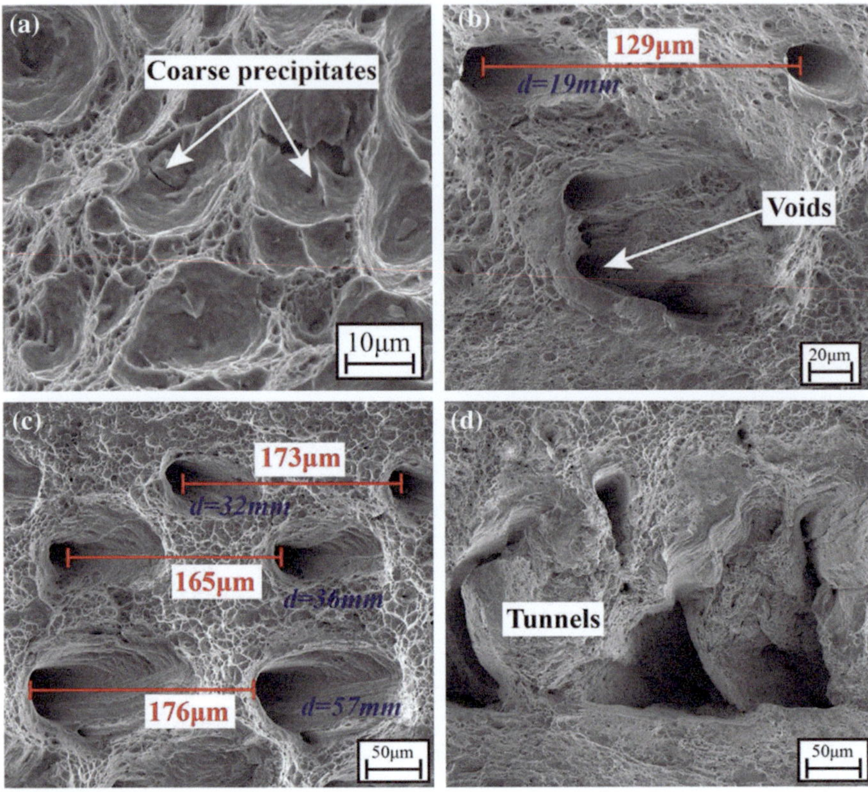

Fig. 3.15 Partly fracture surface morphologies of high depth-to-width ratio FSW: **a** 30 mm/min, **b** 300 mm/min, **c** 400 mm/min and **d** 600 mm/min [23]

their corresponding influence on the properties of the friction stir weld produced of AA2014 aluminium alloy. The results showed that higher microhardness and corrosion resistance were found at lower rotational speed and higher traverse speed. Vijaya Ramnath et al. [25] developed a milling fixture and its clamping setup friction stir welding process conducted on a vertical milling machine. The fixture was designed using three different materials, namely cast iron (C45), die steel (D4) and tool steel. The authors conclude that C45 material can be used as alternate fixture material compared to the other materials investigated. Rzaev et al. [26] developed a mathematical model of temperature dynamics for friction stir welding process. The model was also used to estimate the linear velocity of the FSW process. The model was found to be in satisfactory agreement with the experimental data. Liu et al. [27] investigated the strain rate dependent micro-texture evolution during the material flow for the friction stir welding of pure copper. Simple shear textures were found to be developed at a relatively low strain rate [27]. This phenomenon was attributed to the self-adaption of the crystallographic orientation for the change in the flow stress determined by the change in the strain rate during the material flow. This could be

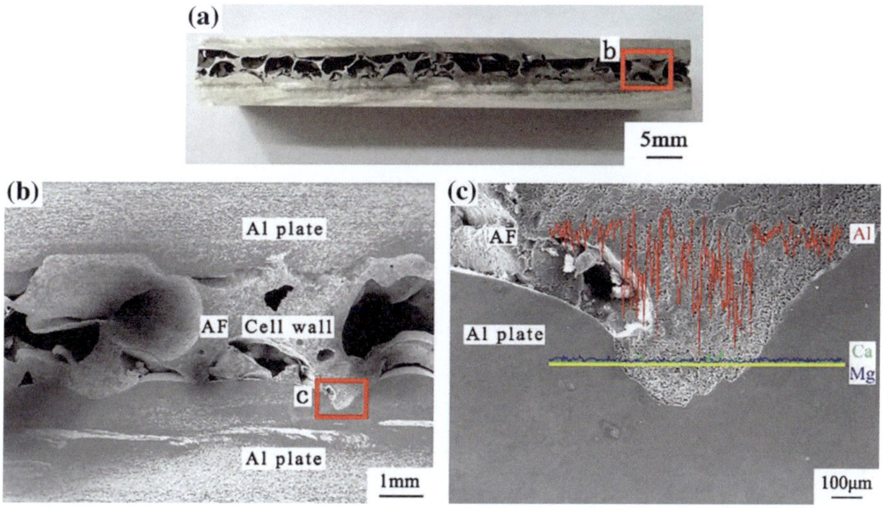

Fig. 3.16 Cross-sectional morphology of AFS: **a** Macro-morphology, **b** microstructure of labelled region (**b**) in **a** and **c** microstructure of labelled region (**c**) in **b** and EDS result [18]

the reason why the various shear textures are observed in different locations within the weld produced in friction stir welding process. Peng et al. [28] used friction stir welding to fabricate aluminium foam sandwich (AFS). The results showed that the joining of the face sheets and aluminium foam was made possible for the plasticized flow of materials. Figure 3.16 shows the cross-sectional morphology of the AFS produced. The AFS produced using the FSW has high flexural strength and high impact resistance with good sound absorption It was concluded that the FSW was a good method for preparing AFS.

External cooling is often employed during friction stir welding process to control the localized properties of the joint. Singh et al. [29] studied the influence of different cooling environments on the properties of the friction stir welded AA5052 strain-hardenable aluminium alloy. The three different cooling environments used in this study was natural air, water and liquid nitrogen. The influence of these cooling media on the metallurgical properties and mechanical properties of the weld produced. The results of this investigation revealed that the width of weld nugget zone, thermo-mechanically affected zone and the heat-affected zone were found to decrease with the respective change in cooling media. The influence of the welding speed on the weld properties was also investigated. The results showed that water cooling produced finer grain size than the grains produced with the liquid nitrogen cooling medium and hence, water cooling gave a better weld quality making it choice cooling medium in friction stir welding. The essence of this cooling was to reduce the extent to which softening of the joint occur, reduce the width of the nugget and heat-affected zones. The grain growth that the welding temperature courses in the heat-affected zone results in grain coarsening which can degrade the material properties in this area. With proper cooling, this grain coarsening can be greatly reduced

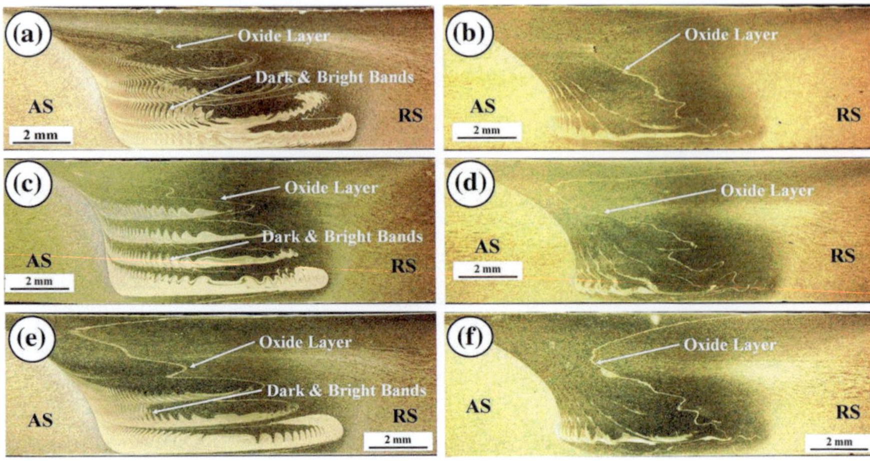

Fig. 3.17 Macrostructure of welds produced in: naturally cooled weld at **a** low and **b** high welding speeds, water-cooled weld at **c** low and **d** high welding speeds, liquid-nitrogen-cooled weld at **e** low and **f** high welding speeds [29]

and the properties improved. Air cooling produces the least influence because of its low effectiveness followed by liquid nitrogen. Increasing the welding speed was found to reduce the heat input during welding which changed the fracture location during tensile testing from the approach side to the retract side of the weld due to considerable change in hardness. The optimum welding condition is low welding speed using water as a cooling medium. These will produce a more ductile weldment and improved properties of the heat-affected zone. Figure 3.17 shows the micrograph of the cross section of friction stir welded joints produced under different welding conditions. The parabolic-shaped cross section of nugget zone was due to the cylindrical tool pin. The well-defined onion rings in the nugget zone is made up of alternative layers of dark and bright bands caused by the mixing of material by tool rotation in weld nugget which are more pronounced as low welding speed with more bands produced in air cooling, with the lowest occurring in water cooling. Also, the higher the welding speed, the lower the band formation. The oxide layer or kissing bond defect observed in the weld nuggets zone due to the oxide entrapment in weld nugget during welding. Krutzlinger et al. [30] demonstrated a method of predicting the morphology of the cross-sectional interfacial area in friction stir welded aluminium/copper lap joints using image processing and Gaussian process regression. A data-driven model of the interfacial area's morphology was developed. The model was tested for the morphological predictions with cross sections welded using the test parameters that were not used for training the developed model. This is to enable to estimate which joining mechanism is dominant for the overall joint strength. The model agrees well with the experimental data. This approach showed that the interfacial area of friction welded joint can be tailored or customized for lap joints. The welding conditions can be tailored to different requirements using the model without the need to conduct expensive experiments and subsequent experimental analysis of the cross sections. The authors concluded that the robustness of the model can be improved by considering additional input data for the training of the model. These include the material specifications, more material combinations or different workpiece dimensions. This would help to expand the existing database for better estimation of joint characteristics. The model can also be applied in real time for online monitoring of the morphology of the interfacial area.

Tejonadha Babu et al. [31] investigated the influence of interchanging the base materials arrangement, that is, advancing side and retreating side on mechanical properties—tensile and yield strength of the weldment using image processing technique. The results of this study showed that the first mode of metal transfer increases as the travel speed was increased which, in turn, have a great influence on the mechanical properties of the joint. Improved mechanical properties were achieved when AS5052 is placed on the advancing side while the RS6061 was placed on the retracting side (AS5052RS6061). In both conditions, defect-free welds were produced. The improved properties were attributed to the proper mixing of the materials in the AS5052RS6061 joints.

Kaid et al. [32] studied the thermo-mechanical behaviour of friction stir welding process of 6061-T6 aluminium alloy. Three dimensional (3D), transient, non-linear structural–thermal model was developed to simulate the distribution of the temperature and the mechanical stresses during the friction stir welding process of the aluminium alloy. The experimental data was compared with the simulated results of temperature distributions (profile and peak temperature) and the residual stress. Therefore, the authors recommended the use of a heat transfer model for the prediction of the temperature distribution during the friction stir welding process. Mironov et al. [33] investigated the effect of welding temperature on the material flow during friction stir welding of AZ31 magnesium alloy. The results revealed that the material motion consists of two major components namely: the shoulder induced flow and the probe induced flow through the entire temperature range and a transition between the two occurring at low temperature. In some cases of welding tool design, as the welding temperature is increased, the contribution of the shoulder and transitional components was found to also increase. At a welding temperature of over 0.8 of melting temperature, Tm, the transition material flow disappear due to reduction in the temperature sensitivity of flow stresses and the resulting equilibration of their distribution within the stir zone. Anbukkarasi and Kailas [34] studied the influence of interlayer on the mechanical properties of AA2024–Cu joints. Zinc was used as the interlayer in this study. The results of this investigation showed that the Zn interlayer was found to diffuse well into the AA2024 matrix at the optimum tool-offset position. The intermetallic compound of Al–Zn–Cu that was seen in the weld was found to increase the strength and elongation percentage when compared with Cu–AA2024 intermetallic. It was concluded that the weld properties can be improved by using appropriate interlayer material. A similar investigation was carried out by AmlanKar and Kailas [35]. The authors studied the influence of zinc interlayer in Al–Ti during the friction stir welding process. The results of this study also showed that the mechanical mixing of Zn, Al, and Ti changed the phase evolution and prevent the formation of the brittle Al_3Ti intermetallic compound which resulted in the substantial improvement in the mechanical properties of the joint. This mechanical mixing of the three materials helped to promote the formation of fine Ti particles instead of as against the formation of Ti flakes.

Muthumanickam et al. [36] studied the influence of processing parameters on mechanical properties and microstructure of friction stir welded AA2195. The tool rotation speed of between 400 and 1000 rpm and the tool travel speed of between 100 and 300 mm/min were investigated. The results of this study revealed that the tool rotation speed and travel speed have great influence on the heat generation in the welds that in turn affect the geometry of the weld stir zone, the evolution of grain size in stir zone, the tensile property of the weld and the hardness of the weld. The combination of low tool rotation speed and high travel speed promote the development of finer grain size. The tensile strength was found to be significantly affected by the tool travel speed than the tool rotation speed. Flash defect is observed at a high rotation speed of 1000 rpm. Also, a tunnel defect was observed in the weld at a lower tool rotation speed of 200 rpm, tunnel defects were observed. The microstructure of the weld stir zone was seen to consist of finer recrystallized grains are observed at a lower tool rotation speed and higher travel speed. The macrographs of samples produced at different process parameters are shown in Fig. 3.18. The micrographs of the nugget zone at different processing conditions are shown in Fig. 3.19. The optimum process parameters that yield the highest weld efficiency of 73% in as-welded condition was obtained at a travel speed of 300 mm/min and rotation speed of 400 rpm.

Du et al. [37] characterized the friction plug welding of 8 mm thick 2219-T87 plate. The investigation revealed that the bonding interface between the base and

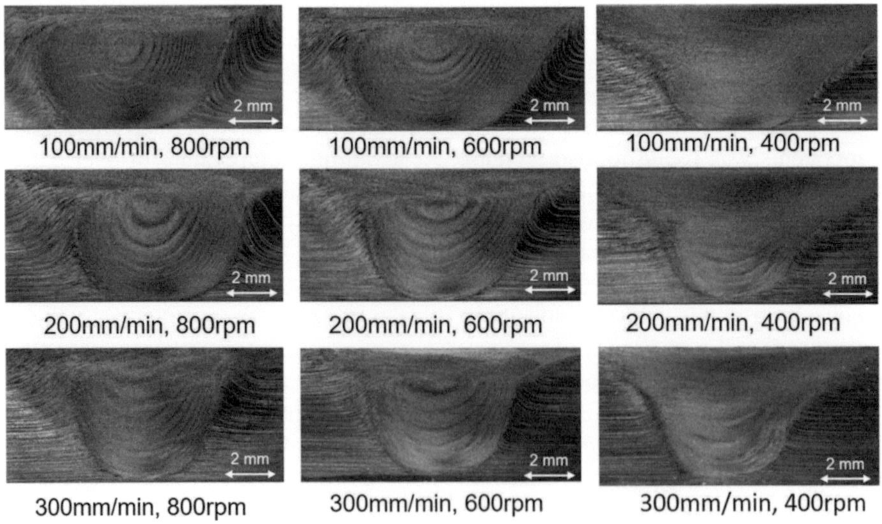

Fig. 3.18 Effect of process parameters on the macrostructure of welds [36]

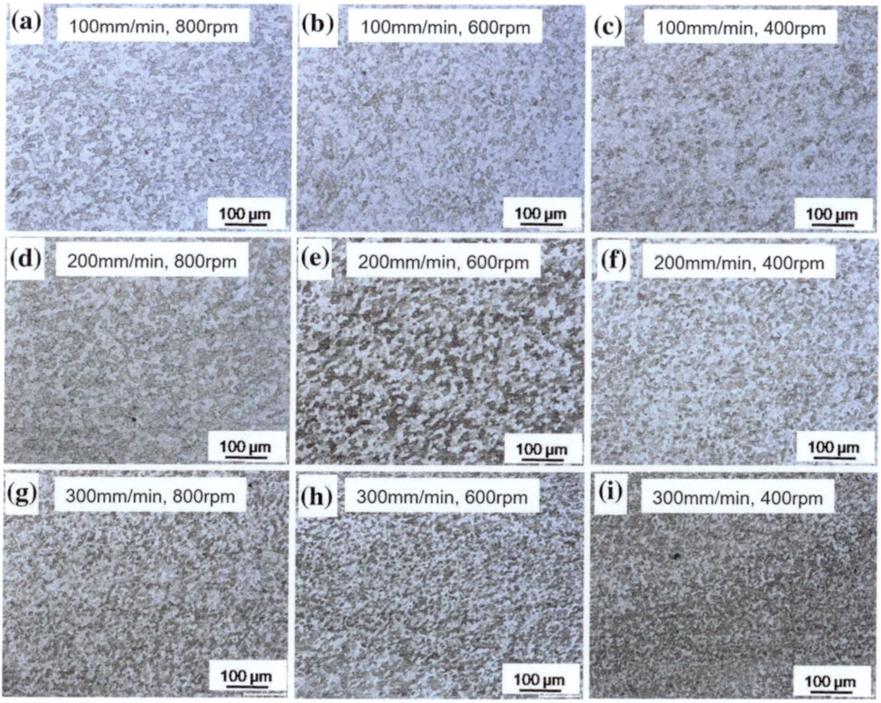

Fig. 3.19 Effect of process parameters on evolution of grains in nugget zone [36]

the plug is characterized by varying width of recrystallized grains as a result of a large amount of heat generated through friction process. The ultimate tensile strength and percentage elongation of the joint produced were greatly improved. Dinda and Ramakrishnan [38] investigated friction stir welding of 12.7-mm-thick high-strength martensitic steel plates. The analysis of the results obtained showed the combination of upper bainite and martensite in the microstructure of the stir zone as shown in Fig. 3.20a–c. Also, the microstructure of the thermo-mechanically affected zone is shown in Fig. 3.20d. The thermo-mechanically affected zone is characterized by fine equiaxed ferrite, bainite, and martensite microstructure. The martensitic structure of the base material was found to be transformed into the tempered martensite in the heat-affected zone due to the heat treatment effect the welding heat has in this zone as shown in Fig. 3.20e. The micrograph of the thermo-mechanically affected zone is also shown in Fig. 3.21a and b showing periodic layers of ferrite, bainite and martensite, and detailed morphology of the ferrite, bainite and martensite microstructure in the TMAZ are shown in Fig. 3.21c and d. There is a general improvement in the mechanical properties of the weld but the hardness obtained in the stir zone is much higher than the other the part of the weld.

Lambiase et al. [39] also carried out an investigation on the effect processing parameters on the metallurgical and mechanical properties of friction stir welding

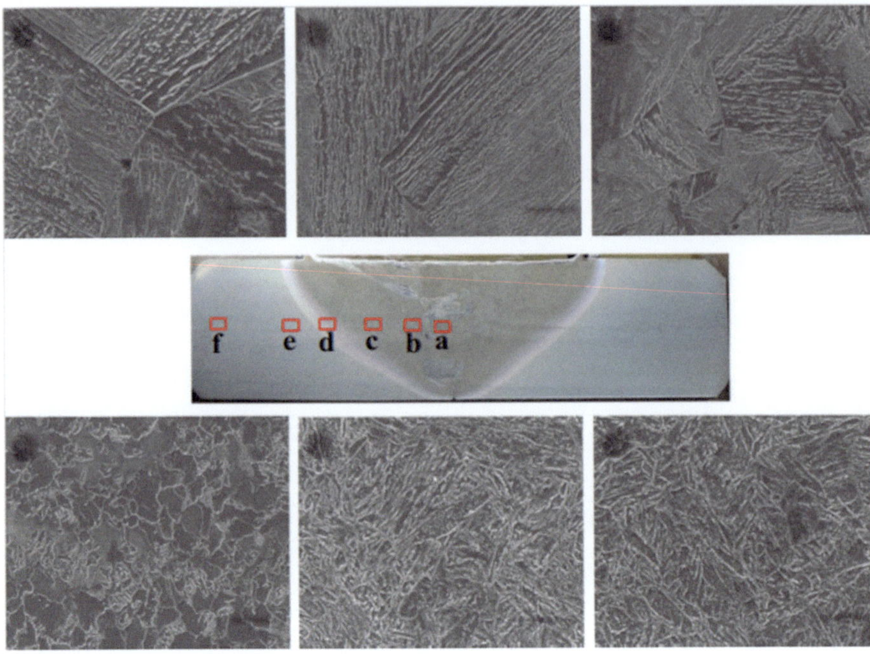

Fig. 3.20 SEM micrographs of **a–c** stir zone, **d** thermo-mechanically affected zone, **e** heat-affected zone and f base material [38]

of polycarbonate performed in butt configuration. The study showed that the welds produced at low welding speed were characterized by an adhesive failure between the stirred region and the substrate. This was as a result of low temperature produced during the welding process. On the other hand, those samples produced at higher welding speed were affected by excessive thinning. The failure of the weld produced at high welding speed occurred in the stirred region or within the base material as a result of localized thinning in these regions. The welding temperature has a direct correlation with the mechanical behaviour of the weld that can be used to control the friction stir welding process. Figures 3.22 and 3.23 show the fracture surfaces of samples produced at low welding temperature and high welding temperature respectively. The fracture surface shown in Fig. 3.22 indicates a better adhesion between the stirred region and the base material and higher elongation (the presence of dimples) as seen at higher magnification in Fig. 3.22b. Figure 3.23 shows specimen failure due to the presence of a sharp notch. The upper weld region shows dimples and fractured bubbles that have undergone high strain while the lower region shows a brittle fracture.

Sekban et al. [40] investigated the microstructural, mechanical, formability and corrosion resistance behaviour of friction stir welded low-carbon steel plates (ASTM 131A) for shipbuilding applications using optimum processing parameters. Two distinct regions, with refined grains separated by a high angle of misorientation and

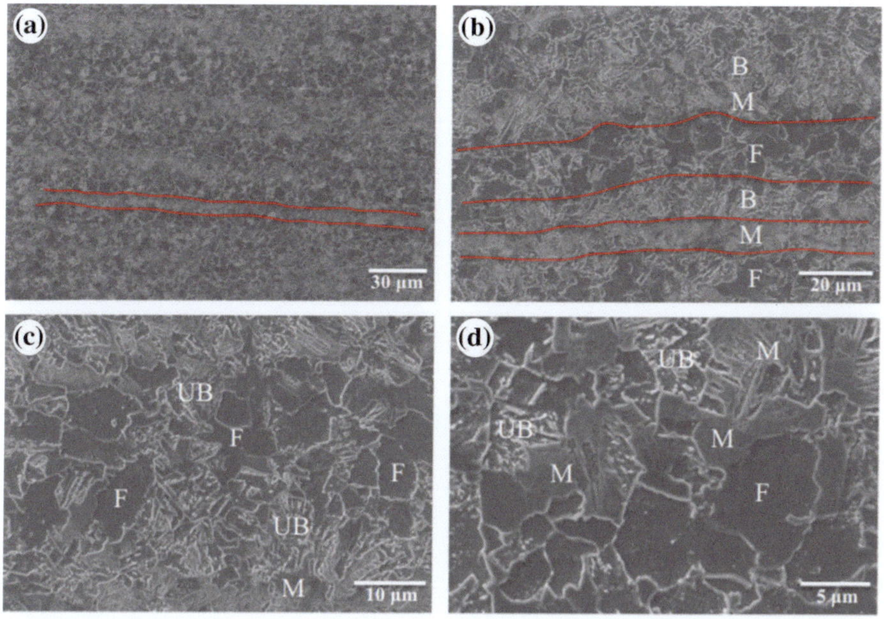

Fig. 3.21 **a** SEM micrographs of the TMAZ showing periodic layers of ferrite, bainite and martensite, and **c–d** showing the detailed morphology of the ferrite, bainite and martensite microstructure in the TMAZ [38]

with the low amount of dislocations were observed in the stir zone and heat-affected zone. The mechanical property, hardness of the stir zone was found to increase from 140 Hv 0.3 to about 230 Hv 0.3 and the yield, and tensile strength were also found to increase from 256 and 435 MPa to about 457 and 585 MPa, respectively, which does not adversely affect the formability of the joint. The corrosion resistance was also found to be slightly increased due to the grain refinement in the weldment. Figure 3.24 shows the microstructure of the uppermost layer (just beneath the shoulder) of the welded zone showing martensitic and bainite grains.

This further confirms the improved mechanical properties of the weld. Figure 3.25 shows the Widmanstätten ferrite and the ferrite–cementite (FC) aggregate structures formed in the stir zone. The Widmanstatten ferritic structure was formed due to intensive plastic deformation and at a low cooling rate with a 10:1 aspect ratio orientation of the ferrite plates with a relatively low dislocation density as shown in Fig. 3.25a. Figure 3.25b shows the TEM micrograph showing the fragmented ferrite–cementite lamellae. Tao et al. [41] also studied the influence of tool travel speed and tool rotation speed on properties of the joint and the fracture behaviour of friction stir welded joints of 3.2-mm thick 2024 aluminium alloys. The results of this study showed that there was no welding defect with the tool travel speed of 100 mm/min and tool rotation speed of between 400 and 600 r/min. Out of these welding parameters, the joint was characterized by tunnel and kissing bond defect caused by precipitate seen

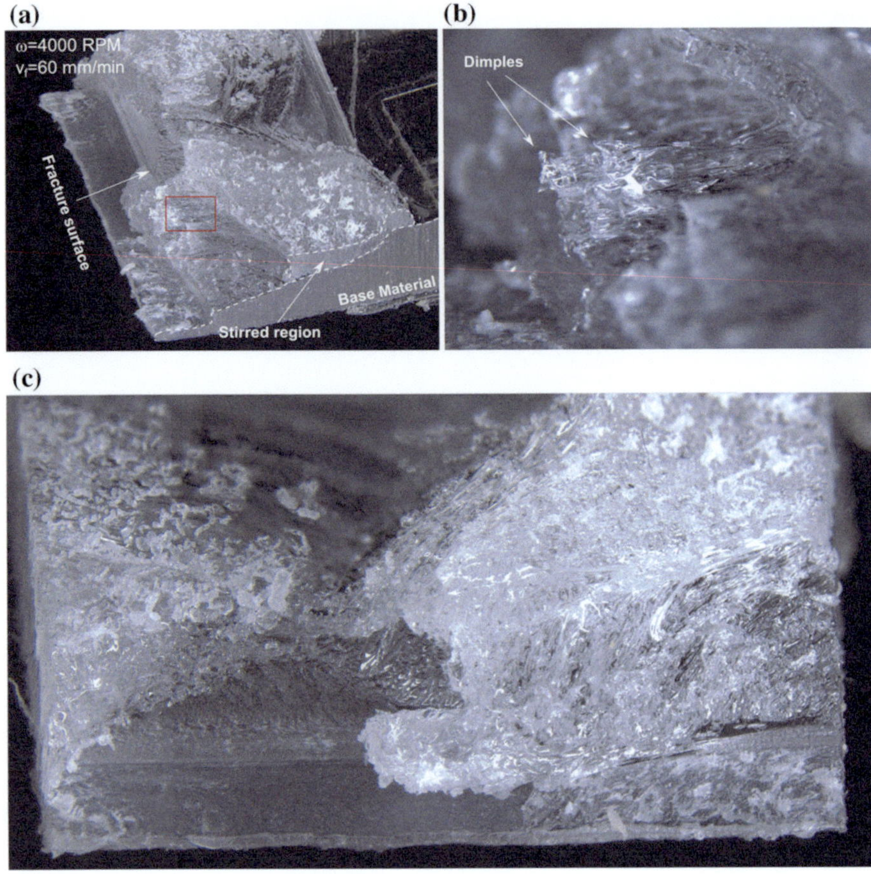

Fig. 3.22 Fracture surface of a sample that failed within the stirred region (rotation speed of ω = 4000 RPM and travel speed of vf = 60 mm/min) [39]

on the bonding line that caused the partial bonding. Verma et al. [42] studied the effect of preheating and water cooling on the thermal, mechanical, metallurgical and texture properties of friction stir welded AA6082. Samples were produced under different preheating (FSW-P) and water cooling (FSW-C) conditions. The results of this study showed that an improved mechanical property was achieved with preheating and water-cooled friction stir welded joint. Preheating was found to have a significant influence on the microhardness and ductility of the joint. The initial temperature of the preheated samples is higher than in water cooling condition with more larger temperature observed on the approach side for all welding conditions. The improved strength was due to proper mixing of material in nugget zone as well as the natural ageing. Qiao et al. [43] explored the friction stir welding of composite plates of 20-mm thick SiCp/2014Al composites. The microstructural and mechanical properties of the composite joint along the transverse and thickness directions

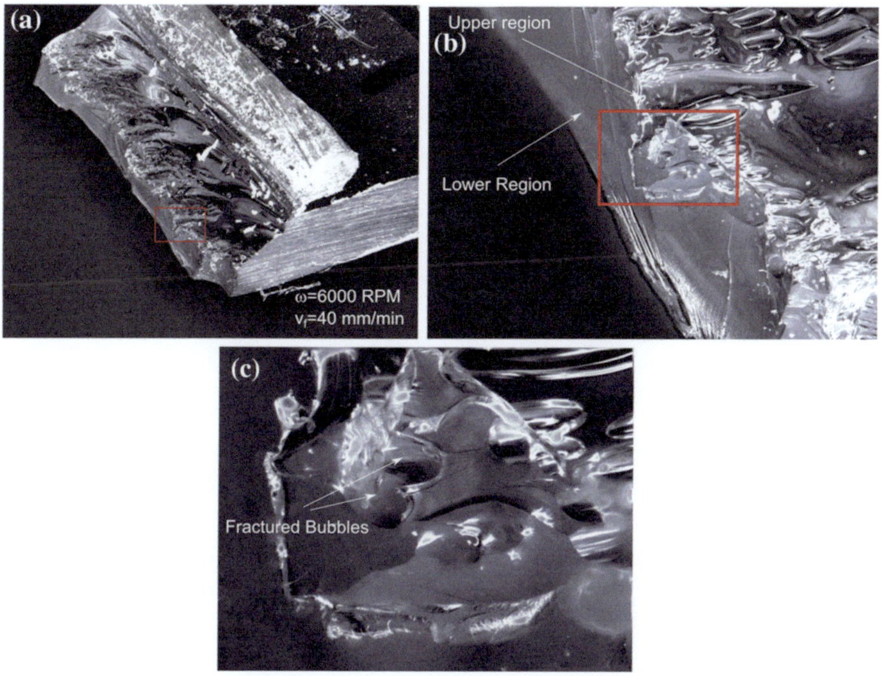

Fig. 3.23 Fracture surface of a sample that failed due to the presence of a sharp notch at the advancing side (ω = 6000 RPM and vf = 60 mm/min) [39]

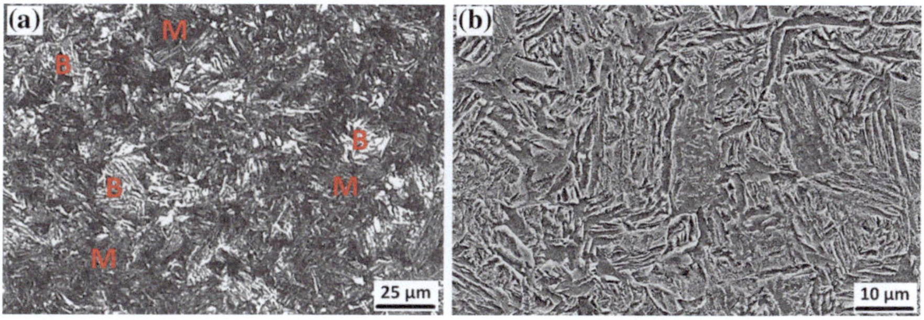

Fig. 3.24 Micrographs showing the martensite and bainite phases of the uppermost layer (just beneath the shoulder) of the welded zone. **a** Optical micrograph and **b** SEM [40]

were studied. The result showed that the microstructure along the transverse direction from the base material (BM) to the heat-affected zone and then to the nugget zone (NZ), uniformly distributed broken down SiC particles are seen with refined grains of aluminium alloy. The Vickers microhardness of the composite joint was also found to gradually increase with the peak values seen at the nugget zone. Li et al. [44] studied the microstructure, microhardness and the tensile properties of

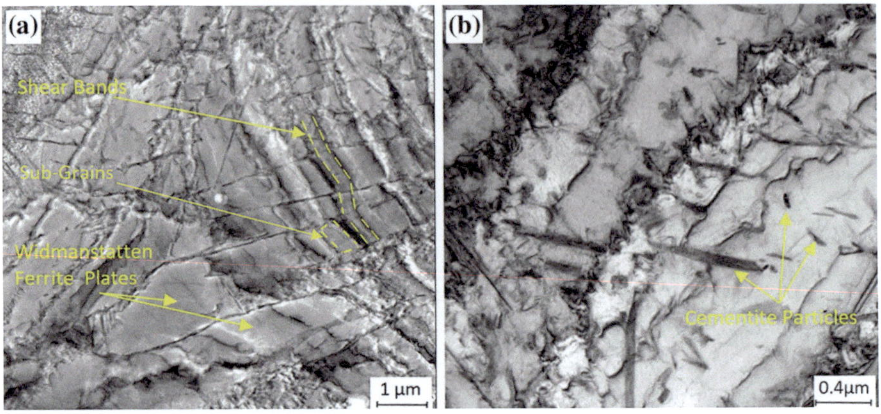

Fig. 3.25 TEM micrographs showing the joint structure of the SZ. **a** Widmanstatten ferrite with shear bands and **b** ferrite–cementite aggregates (FC) with subgrains formed by dislocations [40]

bobbing tool friction stir welded 6082-T6 aluminium alloy with the joints treated by post-weld natural ageing (PWNA) and post-weld artificial ageing (PWAA). The results showed that as the ageing time increased, the microhardness of the stir zone and the thermo-mechanically affected zone (TMAZ) were found to be significantly increased. Also, with increasing PWNA time, the strength and strain were found to be increased slightly while the with increased PWAA time, the strength was increased significantly, The microstructure of the welded joint in the stir zone is characterized by smaller equiaxed recrystallized grains while the grains in the thermo-mechanically affected zone are elongated and equiaxed. Bhushan and Sharma [45] also investigated the microstructure evolution and tensile strength of friction stir welded aluminium alloy (AA6082) composites of AA6082 matrix reinforced with Si_3N_4 and SiC particle. The influence of the rotation speed, tool travel speed and tool tilt angle on the tensile property was investigated. The results showed an improvement in the tensile strength as a result of grain modification of the microstructure during solidification of the composite materials due to strong interfacial bonding between hard SiC particles and AA6082 matrix as well as a homogeneous distribution of SiC particles. Tool rotation speed has the most significant influence of the tensile strength of the weld than the tool travel speed and the tool tilt angle.

A similar study was conducted by Medhi et al. [46] by investigating the influence of the tool traverse speed on the properties of friction stir welded butt joints of 3-mm-thick AA6061-T6 and pure copper. The traverse speed used in this study was 30, 90 and 150 mm/min while keeping the rotational speed at a constant value of 800 rpm. The study revealed the formation of AlCu, Al_2Cu and Al_4Cu_9 phases at low and moderate speeds of 30 mm/min and 90 mm/min while iron aluminide phases (Fe_2Al_5 and $FeAl_3$) and Al–Cu phases were formed at a higher transverse speed of 150 mm/min. The maximum tensile strength was found at a traverse speed 90 mm/min while the lowest joint efficiency and lowest tensile strength were found at the highest traverse speed condition because of the presence of brittle iron–aluminium phases with high-stress concentration factor and hence crack initiation sites. The microstructures of the samples produced at lowest, moderate and highest transverse speeds of 30, 60 and 150 mm/min are, respectively, shown in Figs. 3.26, 3.27 and 3.28.

Tiwari et al. [47] investigated the tool wear in friction stir welding. The authors studied two different grades of tungsten carbide tools: tool A (WC-6 wt% Co) and tool B (WC-10 wt% Co), used in friction stir welding of DH36 steel plates. The results showed that tool A wear morphology was characterized by intergranular failure, which was due to the separations of tungsten carbide grains. The wear mechanism observed in tool B is characterized by adhesion, abrasion, crack initiation,

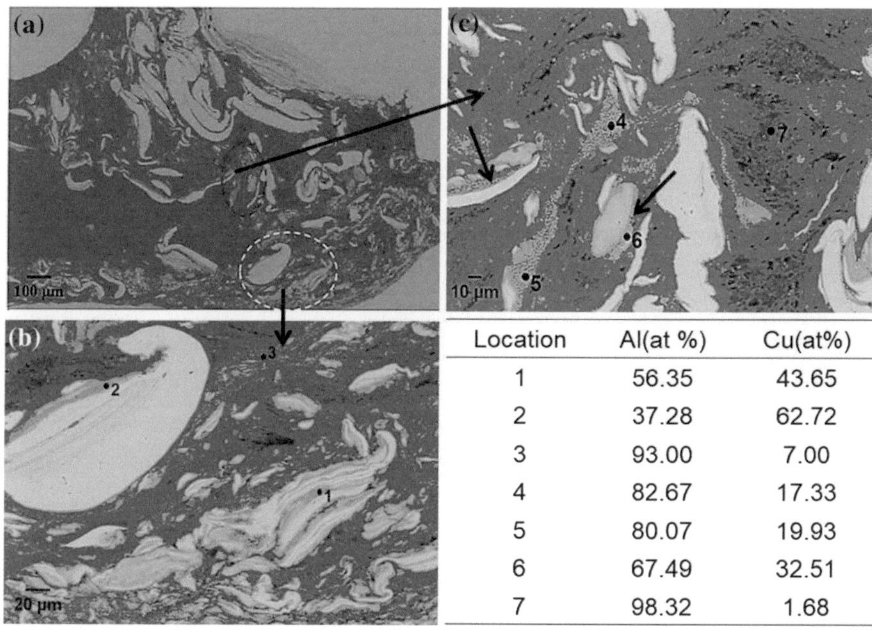

Location	Al(at %)	Cu(at%)
1	56.35	43.65
2	37.28	62.72
3	93.00	7.00
4	82.67	17.33
5	80.07	19.93
6	67.49	32.51
7	98.32	1.68

Fig. 3.26 SEM and EDS analyses at the lowest traverse speed [46]

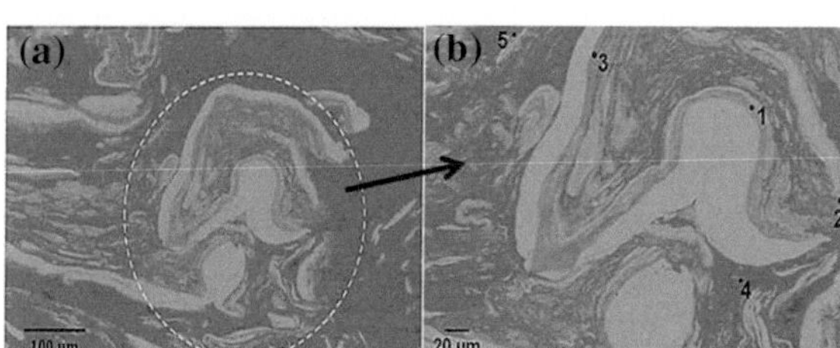

Location	Al(at %)	Cu(at%)
1	42.44	57.56
2	83.30	16.70
3	65.27	34.73
4	98.58	1.42
5	69.73	30.27

Fig. 3.27 SEM and EDS analyses at moderate traverse speed [46]

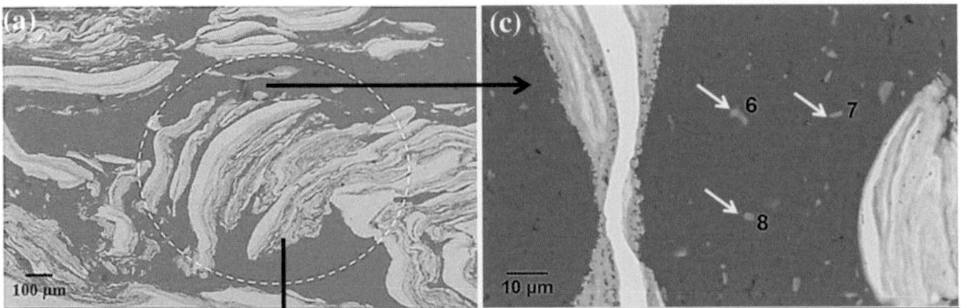

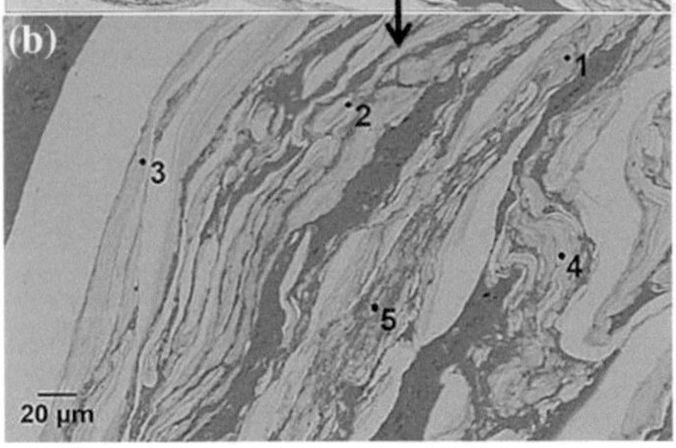

Location	Al(at %)	Cu(at%)	Fe(at%)
1	66.70	33.30	-
2	67.46	32.54	-
3	54.82	45.18	-
4	53.09	46.91	-
5	65.49	34.51	-
6	85.64	0.64	13.72
7	91.88	0.53	7.59
8	89.41	0.58	10.01

Fig. 3.28 SEM and EDS analyses at highest traverse speed [46]

diffusion and oxidation. Sun et al. [48] studied the effect of welding process parameters of friction stir welded dissimilar 2024-T4 and 7075-T6 aluminium alloy on the microstructure and the mechanical properties of the weld. The highest tensile strength of the joints was achieved at a rotational speed of 900 rpm and a transverse speed of 100 mm/min.

3.7 Summary

Friction stir welding process is an important joining technology that is achieved in a solid state of the materials. The problems encountered when joining two dissimilar materials using the traditional welding processes are offset using this welding technique. Different materials can be joined using friction stir welding process. A brief background of this welding process was presented in this chapter. The working principle, advantages and limitations and areas of application are also presented. Friction stir spot welding, another variant of friction stir welding was also presented. Some of the research works to advance the technology was also presented in this chapter.

References

1. S. Lomolino, R. Tovo, J. Dos Santos, On the fatigue behaviour and design curves of friction stir butt-welded Al alloys. Int. J. Fatigue (2005). https://doi.org/10.1016/j.ijfatigue.2004.06.013
2. C.D. Donne, G. Biallas, T. Ghidini, G. Raimbeaux, Effect of weld imperfections and residual stresses on the fatigue crack propagation in friction stir welded joints, in *2nd Friction Stir Welding* (2000)
3. E.T. Akinlabi, S.A. Akinlabi, Effect of heat input on the electrical resistivity of dissimilar friction stir welded joints of aluminium and copper, in *IEEE Africon International Conference*, Livingstone, Zambia, 13–15 September 2011
4. H. Uzun, C. Dalle Donne, A. Argagnotto, T. Ghidini, C. Gambaro, Friction stir welding of dissimilar Al 6013–T4 To X5CrNi18-10 stainless steel. Mater. Des. (2005). https://doi.org/10.1016/j.matdes.2004.04.002
5. M.K. Kulekci, Effects of process parameters on tensile shear strength of friction stir spot welded aluminium alloy (En Aw 5005). Arch. Metall. Mater. **59**(1), 221–224 (2014)
6. Q. Pang, Z.L. Hu, J.S. Song, Preparation and mechanical properties of closed-cell CNTs-reinforced Al composite foams by friction stir welding. Int. J. Adv. Manuf. Technol. **103**, 3125–3136 (2019)
7. T.Y. Pan, Friction stir spot welding (FSSW)—a literature review, in *2007 World Congress, SAE International*, Detroit, Mich (2007)
8. I.J. Ibrahim, G.G. Yapici, Application of a novel friction stir spot welding process on dissimilar aluminium joints. J. Manuf. Process. **35**, 282–288 (2018)
9. M.-J. Hsieh, R.-T. Lee, Y.-C. Chiou, Friction stir spot fusion welding of low-carbon steel to aluminium alloy. J. Mater. Process. Technol. **240**, 118–125 (2017)
10. J.M. Piccini, H.G. Svoboda, Effect of pin length on Friction Stir Spot Welding (FSSW) of dissimilar Aluminum-steel joints. Procedia Mater. Sci. **9**, 504–513 (2015)
11. S. Honggang Dong, Y.S. Chen, X. Guo, X. Zhang, Zhiyang sun refilled friction stir spot welding of aluminium alloy to galvanized steel sheets. Mater. Des. **94**, 457–466 (2016)
12. E. Fereiduni, M. Movahedi, A.H. Kokabi, Aluminum/steel joints made by an alternative friction stir spot welding process. J. Mater. Process. Technol. **224**, 1–10 (2015)
13. M. Movahedi, A.H. Kokabi, S.M. Seyed Reihani, H. Najafi, Mechanical and microstructural characterization of Al-5083/St-12 lap joints made by friction stir welding. Procedia Eng. **10**, 3297–3303 (2011)
14. H. Rao, W. Yuan, H. Badarinarayan, Effect of process parameters on mechanical properties of friction stir spot welded magnesium to aluminium alloys. Mater. Des. (1980–2015) **66**, Part A, 235–245 (2015)
15. R. Ranjana, A.C. de Oliveira Miranda, S.H. Guo, S. Walbridge, A. Gerlich, Fatigue analysis of friction stir welded butt joints under bending and tension load. Eng. Fract. Mech. **206**, 34–45 (2019)
16. B. Sadeghian, A. Taherizadeh, M. Atapour, Simulation of weld morphology during friction stir welding of aluminium stainless steel joint. J. Mater. Process. Technol. **259**, 96–108 (2018)
17. P.H.C.P. da Cunha, G.V.B. Lemos, L. Bergmann, A. Reguly, J.F. dos Santos, R.R. Marinho, M.T.P. Paes, Effect of welding speed on friction stir welds of GL E36 shipbuilding steel. J. Mater. Res. Technol. **8**(1), 1041–1051 (2019)
18. S. Chen, X. Li, X. Jiang, T. Yuan, Y. Hu, The effect of microstructure on the mechanical properties of friction stir welded 5A06 Al Alloy. Mater. Sci. Eng., A **735**, 382–393 (2018)
19. M. Mahmoudiniya, A.H. Kokabi, S. Kheirandish, L.A.I. Kestens, Microstructure and mechanical properties of friction stir welded ferritemartensite DP700 steel. Mater. Sci. Eng. A **737**, 213–222 (2018)
20. R. Bertrand, H. Robe, D. Texiere, Y. Zedan, E. Feulvarch, P. Bocher, Analysis of AA2XXX/AA7XXX friction stir welds. J. Mater. Process. Technol. **271**, 312–324 (2019)
21. W. Tang, X. Yang, S. Li, H. Li, Microstructure and properties of CLAM/316L steel friction stir welded joints. J. Mater. Process. Technol. **271**, 189–201 (2019)

22. E. Gao, X. Zhang, C. Liu, Z. Ma, Numerical simulations on material flow behaviors in whole process of friction stir welding. Trans. Nonferrous Met. Soc. China **28**, 2324–2334 (2018)
23. Y. Huang, Y. Xie, X. Meng, J. Li, L. Zhou, Joint formation mechanism of high depth-to-width ratio friction stir welding. J. Mater. Sci. Technol. **35**, 1261–1269 (2019)
24. S. Sinhmar, D.K. Dwivedi, Effect of weld thermal cycle on metallurgical and corrosion behavior of friction stir weld joint of AA2014 aluminium alloy. J. Manuf. Process. **37**, 305–320 (2019)
25. B. Vijaya Ramnath, C. Elanchezhian, S. Rajesh, S. Jaya Prakash, B. Manoj Kumaar, K. Rajeshkannan, Design and development of milling fixture for friction stir welding. Mater. Today: Proc. **5**, 1832–1838 (2018)
26. R. Rzaev, A. Dzhalmukhambetov, A. Chularis, A. Valisheva, Mathematical modeling of process of the friction stir welding. Mater. Today: Proc. **11**, 591–599 (2019)
27. X.C. Liu, Y.F. Sun, T. Nagira, K. Ushioda, H. Fujii, Strain rate dependent micro-texture evolution in friction stir welding of copper. Materialia **6**, 100302 (2019)
28. P. Peng, K. Wang, W. Wang, L. Huang, K. Qiao, Q. Che, X. Xi, B. Zhang, J. Cai, High-performance aluminium foam sandwich prepared through friction stir welding. Mater. Lett. **236**, 295–298 (2019)
29. R.K.R. Singh, D.W. Rathod, R. Prasad, S. Pandey, Cooling environment effect on the microstructure and mechanical properties of friction stir-welded joints. J. Mater. Eng. Perform. **28**(6), 3737 (June 2019)
30. M. Krutzlinger, E. Meltzer, M. Muehlegg, M.F. Zaeh, Gaussian process regression to predict the morphology of friction-stir-welded aluminum/copper lap joints. Int. J. Adv. Manuf. Technol. **102**, 1839–1852 (2019)
31. K. Tejonadha Babu, S. Muthukumaran, C.H. Bharat Kumar, The role of material location on the first mode of metal transfer and weld formation in dissimilar friction stir welded thin sheets. Trans. Indian Inst. Met. **72**(6), 1589–1592 (2019)
32. M. Kaid, M. Zemri, A. Brahami, S. Zahaf, Effect of friction stir welding (FSW) parameters on the peak temperature and the residual stresses of aluminium alloy 6061-T6: numerical modelisation. Int. J. Interact. Des. Manuf. (IJIDeM) **13**, 797–807 (2019)
33. S. Mironov, Y.S. Sato, H. Kokawa, Influence of welding temperature on material flow during friction stir welding of AZ31 magnesium alloy. Metall. Mater. Trans. A **50**, 2798 (June 2019)
34. R. Anbukkarasi, S.V. Kailas, Role of third material (interlayer) on mechanical properties of the aa2024–copper joints carried out by friction stir welding (FSW). Trans. Indian Inst. Met. **72**(6), 1603–1606 (2019)
35. S.V. AmlanKar, S.S. Kailas, Effect of mechanical mixing in dissimilar friction stir welding of Aluminum to Titanium with Zinc interlayer. Trans. Indian Inst. Met. **72**(6), 1533–1536 (2019)
36. A. Muthumanickam, P. Gandham, S. Dhenuvakonda, Effect of friction stir welding parameters on mechanical properties and microstructure of AA2195 Al–Li alloy welds. Trans. Indian Inst. Met. **72**(6), 1557–1561 (2019)
37. B. Du, X. Yang, K. Liu, Z. Sun, D. Wang, Effects of supporting plate hole and welding force on weld formation and mechanical property of friction plug joints for AA2219-T87 friction stir welds. Weld. World **63**, 989–1000 (2019)
38. G.P. Dinda, A. Ramakrishnan, Friction stir welding of high-strength steel. Int. J. Adv. Manuf. Technol. https://doi.org/10.1007/s00170-019-04003-7
39. F. Lambiase, V. Grossi, A. Paoletti, Advanced mechanical characterization of friction stir welds made on polycarbonate. Int. J. Adv. Manuf. Technol. https://doi.org/10.1007/s00170-019-04006-4
40. D.M. Sekban, S.M. Aktarer, G. Purcek, Friction stir welding of low-carbon shipbuilding steel plates: microstructure, mechanical properties, and corrosion behavior. Metall. Mater. Trans. A. https://doi.org/10.1007/s11661-019-05324-8
41. B. Tao, F. Xia, Y. Zhang, Effect of welding parameters on junction structure and fracture behavior of friction-stir-welded joints of 2024 aluminum alloys. J. Mater. Eng. Perf. https://doi.org/10.1007/s11665-019-04211-y
42. S. Verma, M. Gupta, J.P. Misra, Effect of preheating and water cooling on the performance of friction-stir-welded aviation-grade aluminum alloy joints. J. Mater. Eng. Perform. **28**(7), 4209 (July 2019)

43. Q. Qiao, Y. Su, Q. Ouyang, D. Zhang, X. Song, L. Guo, Microstructural characterization and mechanical properties of 120-mm ultra-thick SiCp/Al composite plates joined by double-sided friction stir welding. Metall. Mater. Trans. A **50**, 3589 (August 2019)
44. Y. Li, D. Sun, W. Gong, L. Liu, Effects of postweld aging on the microstructure and properties of bobbin tool friction stir-welded 6082-T6 aluminium alloy. Int. J. Miner. Metall. Mater. **26**(7), 849 (July 2019)
45. R.K. Bhushan, D. Sharma, Optimization of parameters for maximum tensile strength of friction stir welded AA6082/Si_3N_4 and AA6082/SiC composite joints. Silicon. https://doi.org/10.1007/s12633-019-00216-3
46. T. Medhi, M.K. Yadava, B.S. Roy, S.C. Saha, An experimental investigation on implications of traverse speed in joining of dissimilar Al–Cu by friction stir welding. Int. J. Adv. Manuf. Technol. https://doi.org/10.1007/s00170-019-04086-2
47. A. Tiwari, P. Pankaj, P. Biswas, S.D. Kore, A. Gourav Rao, Tool performance evaluation of friction stir welded shipbuilding grade DH36 steel butt joints. Int. J. Adv. Manuf. Technol. **103**, 1989–2005 (2019). https://doi.org/10.1007/s00170-019-03618-0
48. G.-Q. Sun, G.-S. Xu, D.-G. Shang, & S.-J. Chen, Welding parameter selection and short fatigue crack growth of dissimilar aluminium alloy friction stir welded joint. Weld. World https://doi.org/10.1007/s40194-019-00773-6

Chapter 4
Friction Stir Processing

Friction stir processing (FSP) is variant of friction stir welding (FSW) process, which is a solid-state material processing technique. Materials for specialized applications are desired to have certain selective properties which are not possible to achieve from the traditional alloy processes, such as a combination of hardness and toughness, that can be successfully produced using friction stir processing. Friction stir processing is implemented by plunging a rotating tool on the surface of the material to be processed and stirring of the processing area. Sufficient heat is generated during the stirring and translation movement of the tool across the material surface to cause plastic deformation, creating a dynamic recrystallization, along the processing zone. The resultant microstructure from FSP is ultrafine grains with modified properties. In this chapter, friction stir process is presented with the brief background of the process, the principle of operation of this technique, advantages, limitations and areas of application. Some of the current research work in this field are also presented in this chapter.

4.1 Introduction

Friction stir processing (FSP) is a variant of friction stir welding process that was discussed in Chap. 3 of this book. Friction stir processing is a solid-state processing technique that is used to improve properties of the material such as toughness, wear resistance and fatigue. The friction that is generated between the tool and workpieces causes an increase in temperature that softens the workpiece which causes the workpiece to become extensively plasticized. This excessive plastic deformation leads to significant grain refinement that in turn changes the property of the material. During FSP, the grains of the material under processing becomes homogeneous, ultrafine and equiaxed which help to improve the properties of the material that can help to improve other manufacturing processes performed on the material such as making

E. T. Akinlabi and R. M. Mahamood, *Solid-State Welding: Friction and Friction Stir Welding Processes*, Mechanical Engineering Series,
https://doi.org/10.1007/978-3-030-37015-2_4

metallic alloys bendable more than it can be achieved without the FSP. Also, certain applications demand special surface characteristics to improve for instance wear resistance and corrosion resistance properties, FSP is an important material processing technique that can help to achieve these objectives. Friction stir processing was developed by Mishra et al. [1, 2]. The authors used this FSP to modify the microstructure of material for property enhancement that is based on the principles of friction stir welding process. FSP is a rapidly growing material processing technique that found its application in fabrication, processing, and synthesis of metallic materials [3–5].

In this chapter, the principle of operation of FSP is discussed. The processing parameters that influence the properties of the material in FSP are also highlighted. The advantages, limitations and areas of application of FSP are presented. The importance of the material processing technique in various applications has made this technique to be of great interest in the research community. Some of the research works that have been reported in the academic literature on FSP are also presented in this chapter.

4.2 Working Principle Friction Stir Processing

Friction stir processing (FSP) uses the working principle of Friction stir welding to transform a heterogeneous microstructure to a more homogeneous and a more refined microstructure. The friction stir processing uses the FSW tool to stir up the microstructure of the material surface and breakdown the existing grains and grain boundaries thereby producing a uniform size grains or equiaxed grains. The maximum temperature attained during the FSP process is far less than the melting point of the metal which prevents the formation of grain structure that is produced through equilibrium solidification that is typical of a liquid metal. This processed surface has superior strength and improved formability than the base material, Friction stir processing is a thermo-mechanical material processing technique, the heat in the process is generated through the mechanical stirring and the whole process is a solid-state process that material melting does not occur. It modifies the microstructure of the material unlike the FSW that is used in the joining of materials.

The rotating nonconsumable cylindrical tool pin is plunged into the material surface. The tool pin has a smaller diameter than the diameter of the shoulder that controls the penetration depth of the tool as shown in the diagram in Fig. 4.1. When the rotating tool descends on the surface of the part, the friction between the tool and the workpiece produced heat that softens and plasticize the metal while the tool transports the metal from the leading face of the pin to its trailing edge. The contact of the rotating shoulder of the tool with the outer part of the material column also results in the generation of more heat that further plasticizes the metal around the inserted pin and it also provides an additional forging force that is used to contain the upward flow of material produced by the tool pin. To process the entire surface of the material, a series of overlapping passes of the tool in relative motion with the material is made in order to cover the entire surface. The material undergoes intensive plastic deformation, rigorous mixing as well as a thermal treatment due to

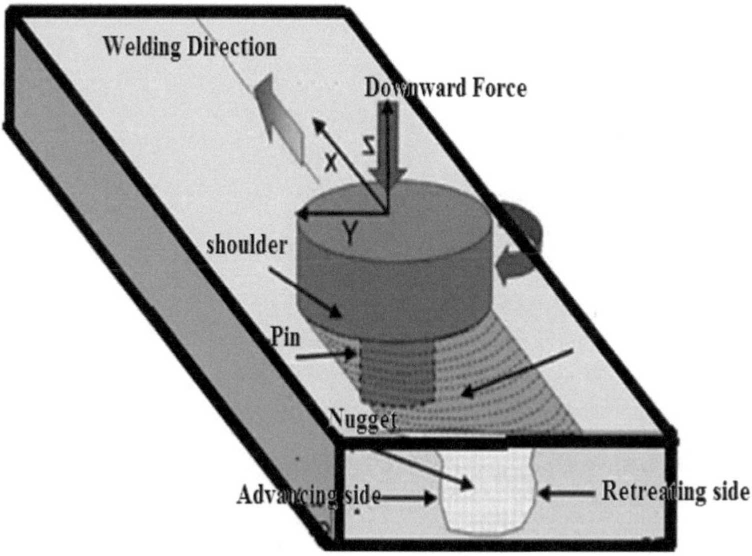

Fig. 4.1 Schematic diagram of FSW

increased temperature from friction and mixing actions. The resulting microstructure is a defect free, recystallized, equiaxed fine grains and uniformly distributed second phase grains that have a great impact on the properties of the processed material. The friction stir processing is also used to produce composite material where the reinforcing material is stirred into the base material thereby improving the properties such as wear and corrosion resistance of the base. The main functions of the tool in the FSP are to stir the material, produce localized heating and control the material flow. The tool shoulder helps in the control of material flow from the displaced material by the tool pin. The tool pin is provided with a small thread that allows the tool to be drilled into the material.

4.3 Processing Parameters in Friction Stir Processing

The most important processing parameters in friction stir processing are the rotational speed, the travel speed, tool geometry design and downward and normal force. These processing parameters have a great effect on the microstructure of the processed material. These processing parameters are described below:

4.3.1 Tool Rotational Speed

How fast the tool rotates has an influence in the type of microstructure development in the friction stir processing which in turn affects the properties of the processed material. The tool rotational speed affects the degree of grain refinement. If a higher temperature is desired, the rotational speed needs to be higher. The rotational speed can be controlled in order to control the microstructure of the material. The rotational speed is also an important processing parameter that influences the degree of plasticity and flow of material achievable in the processed material.

4.3.2 Traverse Speeds

This is the travel speed of the tool across the surface of the material. The combination of the tool transverse and the tool rotational speed has a great effect on the evolving properties of the material being processed. A typical combination of transverse speed and rotational speed with lower rotational speed and higher traverse speed produced a higher microhardness in the processed material [6]. The right combination of these two processing parameters is needed to be optimized in order to achieve the required microstructure and hence the desired properties.

4.3.3 Tool Geometry Design

Tool geometry is another important process parameter in friction stir processing. Researchers have tried various tool geometries and they found out that they have a great effect on the resulting properties in friction stir processing [7, 8]. cylindrical tool, square tool, threaded tool and pinless (see Figs. 4.2 and 4.3) tool have been used in friction stir processing with a variety of properties resulting from their use. Depending on the microstructure desired from the material being processed, any of these tool geometry can be used.

4.3.4 Downward or Normal Force

The downward forces can have a great influence on the microstructural development in friction stir processing. This is the normal force that is applied to the rotating tool. This downward force is responsible for creating the required friction between the tool and the material being processed that in turn results in the generation of heat that is used in the friction stir process. There is a need to properly control this important friction stir processing so as to have a consistent downward force that is required to produce the desired microstructures as well as the desired properties [9].

Friction stir processing has a number of advantages and it is used in a number of industries. The advantages, limitations and areas of application are presented in the section.

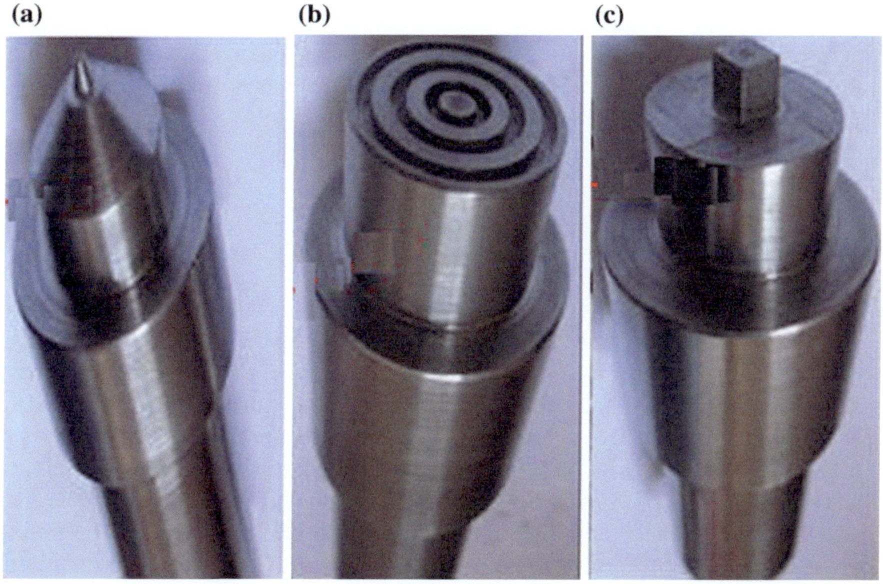

Fig. 4.2 Tool geometry **a** cylindrical tool **b** pinless tool **c** square tool [7]

Fig. 4.3 Tool with threaded pin [8]

4.4 Advantages, Limitations and Areas of Application of FSP

There is always a constant need for materials for specialized applications possessing specific properties. This is an important task in many industrial applications such as in aerospace and automobile industries Materials need to be designed in order to have material with the desired properties. Often applications may require materials with confliction properties such as high ductility and high hardness, these properties cannot exist simultaneously in a metal without specially designing such material. It is possible to have a material with high strength and high ductility if the material has homogenous and ultrafine grain structures. Microstructural modification is a process of altering the initial microstructure of the material and be tailored to the required structure. Friction stir processing has been shown to be beneficial in microstructural modification for various applications. Some of the advantages, limitations and areas of applications of FSP are presented in this section.

4.4.1 Advantages of Friction Stir Processing

- Friction stir welding helps to improve the forming and formability of lightweight alloys at room temperatures that present a serious challenge. One of the ways to improve the formability of such lightweight alloys is through refinement and homogenization of the microstructure in combination with a technique of forming at high temperatures such as superplastic forming. Such conventional grain refinement technique such as hot rolling is very expensive with high energy consumption and it is also time consuming. Friction stir processing has an advantage over other grain refinements techniques. FSP can be used to effectively produce ultrafine grain and homogenized structure at a very low cost with low energy consumption.
- Another advantage of FSP is that it is a single step process making it cheaper and less time consuming when compared to other grain refinements techniques that use multiple steps.
- Friction stir processing also uses an inexpensive and simple tool. Friction stir processing can be achieved on the readily available machine such as a milling machine.
- Friction stir processing can be easily automated and it does not require the use of harmful chemicals and there is no release of harmful gases making it an environmental-friendly process.

4.4.2 Limitations of Friction Stir Processing

The ease of automation, low energy consumption and environmental friendliness are some of the attractive features of friction stir processing, however, there are also some limitations of this important grain refinement technique that includes the following:

- No availability of data is a major limitation of FSP because the process is relatively new. As more and more data become available through more and more research, this limitation will be removed.
- Lack of adequate predictive models for the resulting microstructure is another limitation of this process.
- Another important limitation of friction stir processing is the presence of keyhole at the end of each pass.

4.4.3 Areas of Application of Friction Stir Processing

Friction stir processed zone is characterized by recrystallized, ultrafine grains as well as uniformly distributed second phase particles making it an important grain modification process in many applications where microstructural modification is required. Some of these application areas include the following:

- For locally eliminating casting defects—Casting of metallic parts is a relatively cheaper manufacturing process but with defects such as porosity and other microstructural defects. Friction stir processing can be used to eliminate the casting defect through the vigorous stirring of cast metal to remove porosity, and grain refinement and homogenization to improve properties of the cast metal.
- For fabrication of metal matrix composites (MMC)—Aerospace and automobile industries are in constant need of metal reinforced with ceramic to improve properties such as wear resistance properties. The problem with MMC is the reduction of ductility and toughness because of the incorporation of hard ceramic phases, which may pose difficulty in forming this metal. To overcome this problem, the introduction of the ceramic phases are done on the surface layer of components after it has been formed into the required shape. Friction stir processing can be used to incorporate the ceramic phase on the surface of the part. This FSP technique, being a solid-state process, makes it possible to achieve this objective which is not possible in the past with the traditional techniques that are based on liquid-phase processing. The interfacial reaction between the reinforcement and metal matrix in the liquid state is detrimental to the properties of the composite are prevented using the FSP. Friction stir processing is the best suitable technique for fabricating surface composites on the part. FSP can also be used to produce a nanocomposite in metals to increase the properties such as hardness, and superplasticity in the metal for high resistance to deformation under high-stress conditions.
- Powder metallurgy—Friction stir processing can be used to improve the properties of part produced using powder metallurgy technique. The oxide films on the surface of powder particles are detrimental to the ductility, fatigue, and fracture toughness of the part. Friction stir processing can be used where localized treatment is desired such as in powder metallurgical part.
- Other areas of application of friction stir processing include where improved strength and ductility are desired, where increasing corrosion resistance, improving the fatigue behaviour for enhancing formability, where fine grain microstructure, with superplasticity such as in aerospace and automobile industries.

4.5 Research Advancements in FSP

The importance of friction stir process in surface modification has generated a lot of interest in the research community with constant search to have a better understanding of the process and then to further improve the process for material surface engineering. Some of these research works are presented in this section.

The processing parameters in friction stir processing are found to have a great influence on the evolving properties of the friction stir processed materials. Izadi et al. [6] investigated the influence of tool shape and rotational speed on microstructure and mechanical properties of friction stir processed wrought aluminium alloy 5059. The authors studied three different tool shapes and three different rotation speeds. The influence of these processing parameters on recrystallization, grain growth and mechanical properties (microhardness and tensile strength) in the processed sample was investigated. The three types of tool shapes that were studied are: the grooved pin, the 3-flat threaded pin and the conventional threaded pin. The rotation speeds studied are 638 and 895 RPM. The micrographs of the processed samples with the three tool shapes are shown in Fig. 4.4. The results showed that the shape of the tool pin profile has great influence on the flow of plasticized material and the shape and properties of the stir zone. The grain size was found to be lowest for the 3-flat tool as compared to the largest obtained for the grooved tool. The rotation speed also has a great influence to change the stir zone grain size. The higher the rotation speed, the higher the temperature of the stir zone, which in turn produces larger grain sizes. The 3-flat threaded tool at a rotation speed of 638 RPM generated the finest grain structures. The high grain refinement achieved resulted in improved mechanical properties, the high hardness and high tensile properties.

Wang et al. [10] studied the development of heterogeneous bimodal grains of AlxCoCrFeNi alloy using friction stir processing of Al0.1-CoCrFeNi and Al powders. It was observed that the addition of Al powder resulted in the formation of bimodal-grained structure and second phases that helped to increase strength without compromising the ductility while the result was better than for the friction stir processed base alloy. The addition of Al powders caused some localized Al-rich regions that nucleated the body-centered cubic phase and Al-rich precipitates, that further help to refine the grain thereby increasing the strength. Figure 4.5 showed the formation of the heterogeneous microstructure on the horizontal section of the friction stir processed sample around the exit hole. During the friction stir process, the La powder became melted as a result of the heat generated during the process.

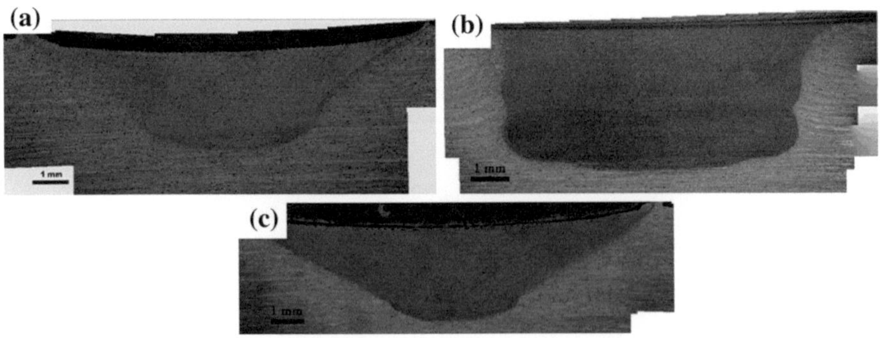

Fig. 4.4 Macrographs showing the stir zone shape of samples friction stir processed with **a** the grooved pin, **b** the 3-flat threaded pin and **c** the conventional threaded pin [6]

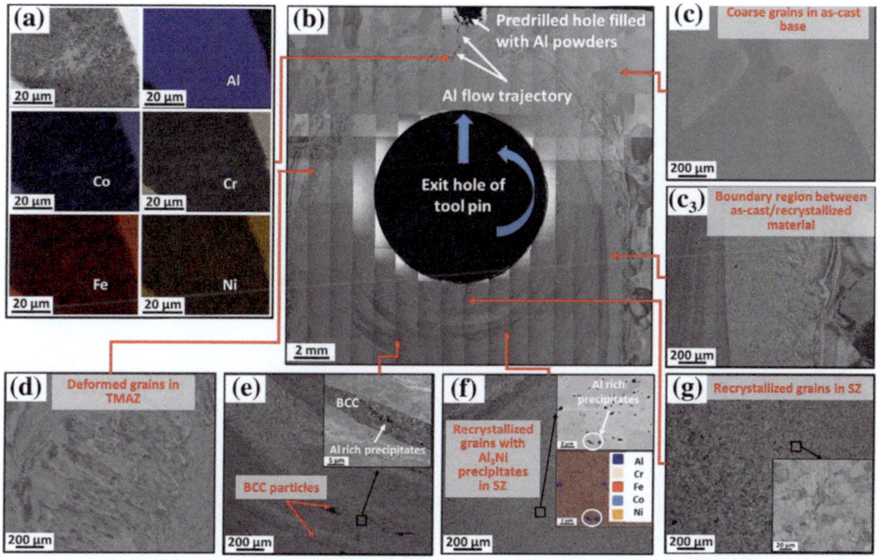

Fig. 4.5 **a** Horizontal section of friction stir processed sample at the exit hole, **b** coarse grains in the as-cast base alloy, **c** boundary region between as-cast/recrystallized zone showing gradient-grained structure, **d** recrystallized grains without Al-rich precipitates in the stir zone (SZ), **e** recrystallized grains with Al-rich precipitates in the SZ, **f** band regions with BCC phase and Al-rich precipitates in the SZ, **g** deformed grains in thermally affected zone (TMAZ) and **h** EDS mapping on the Al flow trajectory showing that Al melted and the other elements reacted with it (extracted from [10])

The melted Al powder was also mixed with the base material as a result of the stirring action of the tool resulted in reactions and phases development in the processed region. The region is made up of a combination of fine grain structures of Al-rich precipitates with finer FCC phase, and recrystallized Al0.1CoCrFeNi Al-rich BCC grains and second phases leading to the heterogeneous bimodal-grain in the microstructure.

Tool pin eccentricities are found to have influence on the material flow as well as the degree of grain refinement in friction stir processing. Chen et al. [11] investigated the influence of tools eccentricities in friction stir processing of Al-5052 alloy on properties development in the alloy. Different pin eccentricities of the tool were investigated. The results obtained from samples processed with tool pin eccentricities are compared to those processed without tool pin eccentricities. The results showed that the pin eccentricities enhances the flow of material and refines grains of the stir zone, which in turn resulted in better yield strength and hardness. However, the pin eccentricity needs to be optimized. An oversized pin eccentricity results in heat generation during the friction stir process that results in the coarsening of grains that deteriorated the mechanical properties. The microstructures of the stir zone in samples processed at different stir tools are shown in Fig. 4.6.

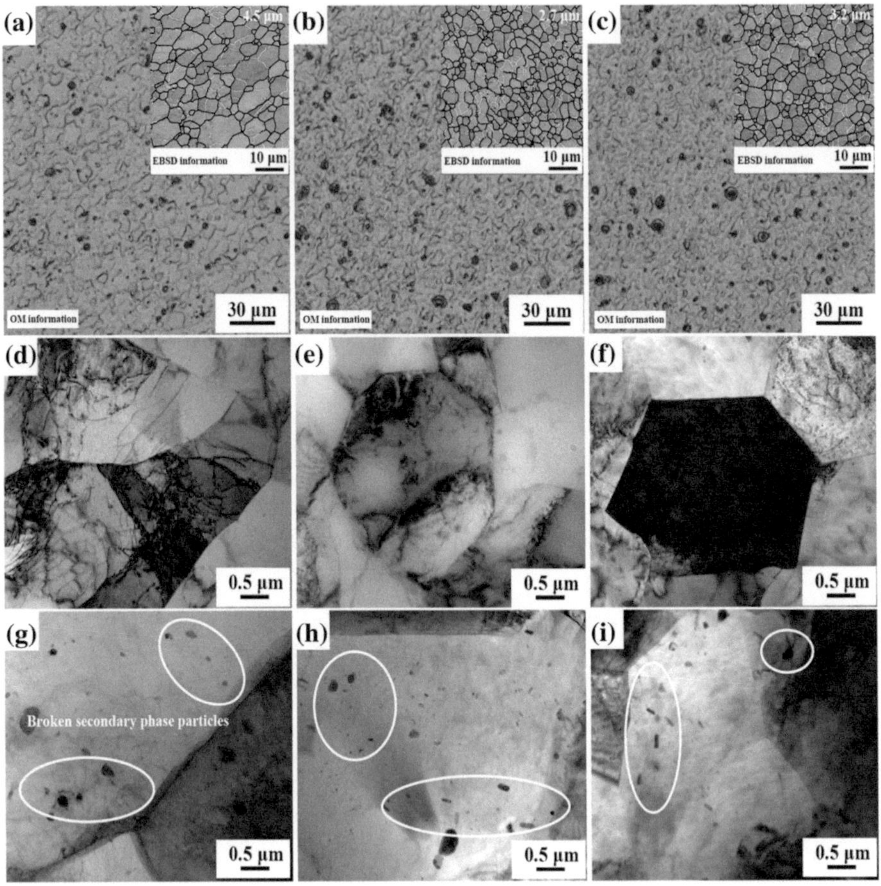

Fig. 4.6 The microstructures in the stir zone produced by different stir tools without eccentricity (**a**, **d**, **g**), with pin eccentricity of 0.4 mm (**b**, **e**, **h**) and with pin eccentricity of 0.8 mm (**c**, **f**, **i**) [11]

The microstructure of the parent material is characterized by coarse grains while the microstructure of the processed samples are dominated by recrystallized fine grains. The microstructures of the samples that are processed with tool eccentricities of 0, 0.4 and 0.8 mm with average grain sizes of 4.5 μm, 2.7 μm and 3.2 μm, respectively, and are shown in Figs. 4.6a–c. The grains become refined and were found to increase as the pin eccentricity is increased from 0 to 0.4 mm while the grains produced at tool eccentricity of 0.8 mm are larger. The reason for this is that, at larger tool eccentricity such as 0.8 mm, the friction produced is larger which causes higher heat generation with slower cooling rate and hence larger grains. Figure 4.6d, f are the TEM images of the SZ of 4.6a–c, respectively. Also, Fig. 4.6g, h are TEM images of Figs. 4.6a, b, respectively.

Friction stir processing has been established as one of the material processing technique for improving the corrosion resistance properties of the processed material. Zhang et al. [12] studied influence of rotation speed on the corrosion property of friction stir processed rolled Mg–Nd–Zn alloy (NZ20). The results showed grain refinement, a homogeneous grain size distribution, less second phases and evolution of stronger basal plane textures. These microstructure resulted in an improved corrosion resistance property. The sample processed at 400 rpm was found to have a better corrosion resistance property when compared to sample produced at 600 rpm. The improved corrosion resistance property in sample processed at 400 rpm was attributed to the increase of basal plane intensity and grain refinement. The hardness was lowered slightly with increased elongation, which might be attributed to the crushing and redistribution of the second phases by the stirring action of the tool. The microstructure of the unrolled NZ20, rolled NZ20, friction stir processed sample at 400 rpm and at 600 rpm are shown, respectively, in Fig. 4.7a–e.

A similar study was conducted by Jiang et al. [13]. The corrosion behaviour of friction stir processing of AISI 304/304L austenitic stainless steel plates at constant tool temperature. The friction stir processing was conducted at varying tool temperature that was achieved through the use of developed temperature control algorithm. The cavitation erosion behaviour of the processed sample was studied and compared to that of the parent material. Tool temperature was found to have an influence on the microstructure as well as the cavitation erosion behaviour in friction stir processing. Friction stir processed samples have higher improvement in cavitation corrosion resistance than the unprocessed sample. The lower average grain size was produced at lower tool temperature than what was obtained at higher tool temperature. A 50% increase in hardness was also obtained in the processed zone, which was attributed to the grain refinement observed. The study concluded that if the heat input and process temperature during the friction stir processing can be controlled, then the cavitation erosion resistance can be improved.

Deore et al. [14] studied the influence of three filler materials on properties of friction stir processing of aluminium alloy AA 7075 surface reinforced composites using carbon nanotubes (MWCNT), Copper (Cu) and Silicon carbide (SiC). The friction stir processing was conducted on the surface after dispersing the filler material and subsequent age hardening. The result of this study showed grain refinement in all the samples. The age hardening further resulted in improvement of properties due to evenly distributed precipitations of the filler materials. The SiC filler was found to give the best properties amongst the three filler materials that were used for the set of friction stir processing and age hardening parameters used in this study. The microhardness, impact toughness and wear resistance was achieved in age hardened friction stir processed SiC filler surface composite. Figure 4.8 shows the microstructure of the base material and those of the three fillet materials after processing.

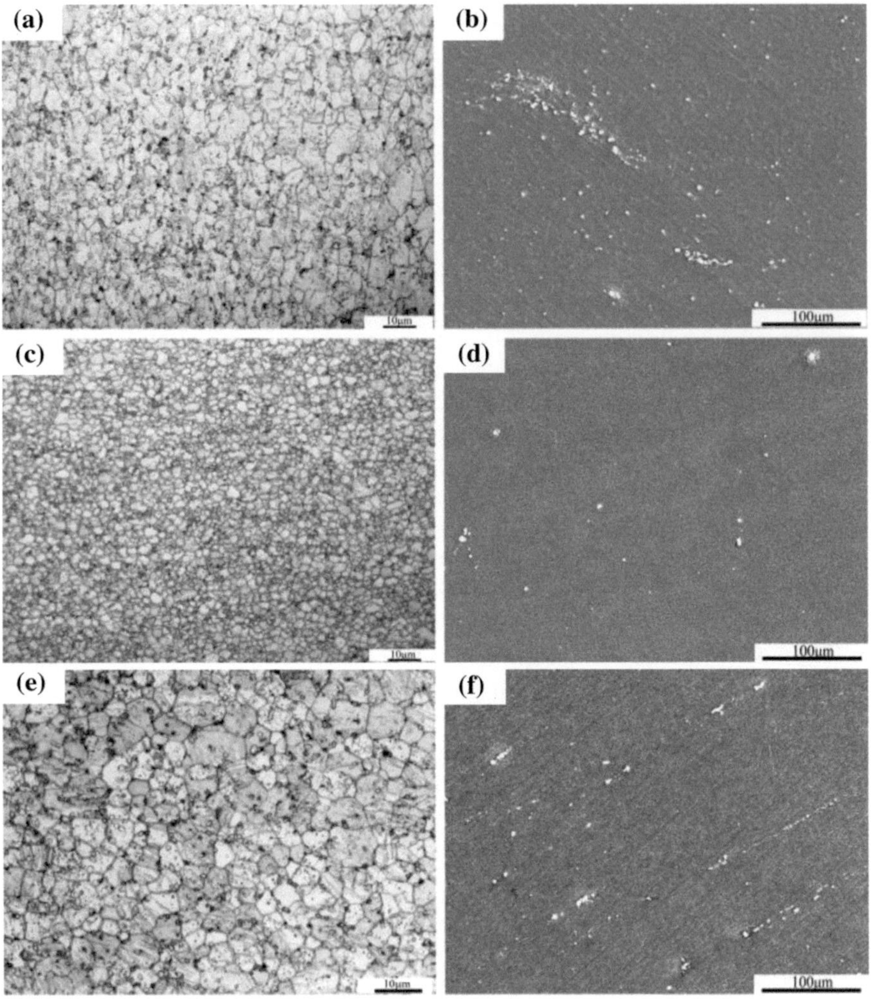

Fig. 4.7 Optical micrographs showing the microstructures of NZ20: **a** rolled sheet; **c** FS-400 rpm; **e** FS-600 rpm, and granular discrete second phases in SEM micrographs of NZ20: **b** rolled sheet; **d** FS-400 rpm; **f** FS-600 rpm [12]

Friction stir processing conducted after friction stir welding has been reported to improve the properties of the friction stir welded joint [15]. Shang et al. [15] successfully introduced profuse extension twins to the stir zone of friction stir welded AZ31 alloy through friction stir processing. The friction stir welded joint was made into a two-layered structure after an additional pass of friction stir processing using a different tool. The resulting texture and the lamellae grain structure has a strengthening effect that significantly improved the mechanical properties of the joint. The

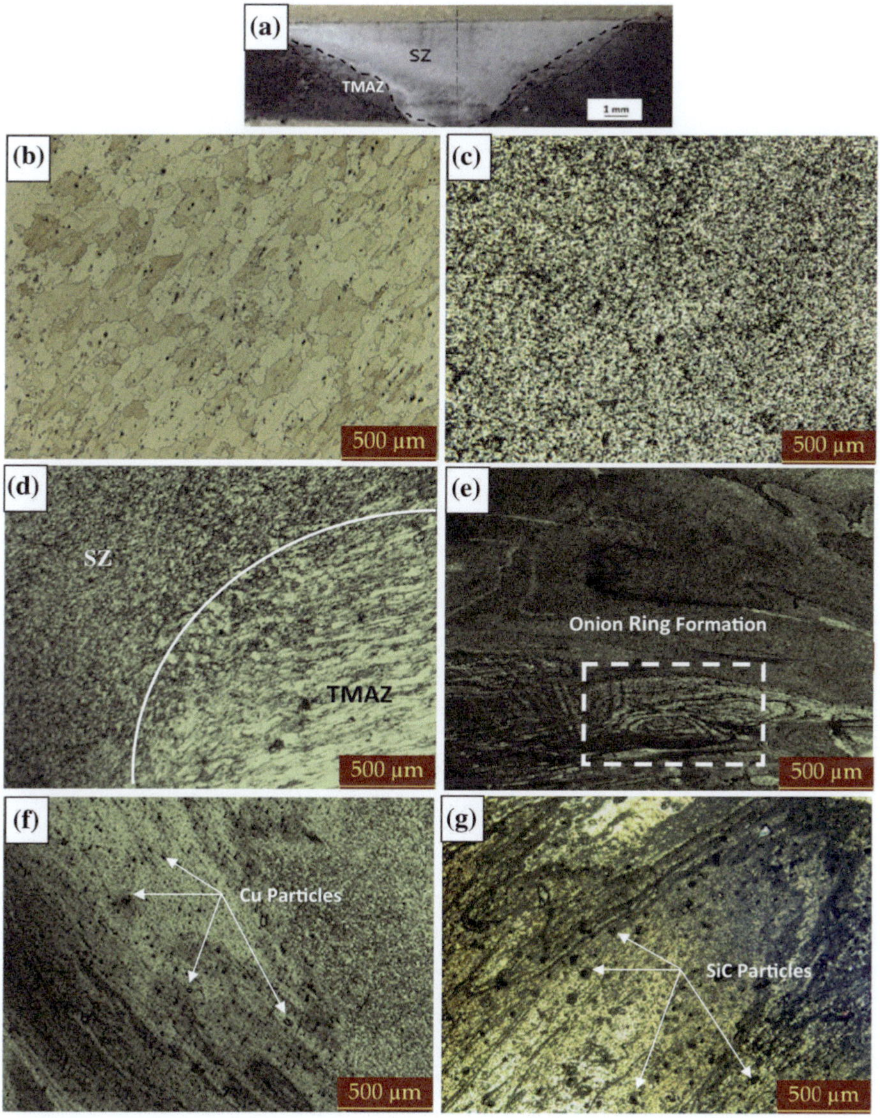

Fig. 4.8 **a** Macrostructure after friction stir processing, Optical Microstructures showing **b** Base Metal AA7075 T6, **c** Stir Zone of friction stir processed (FSPed) AA7075 without filler, **d** Transition between FSPed stir zone and thermo-mechanically affected zone (TMAZ), **e** Stir Zone of FSPed AA7075 with MWCNT filler **f** Stir Zone of FSPed AA7075 with Cu filler, **g** Stir Zone of FSPed AA7075 with SiC filler [14]

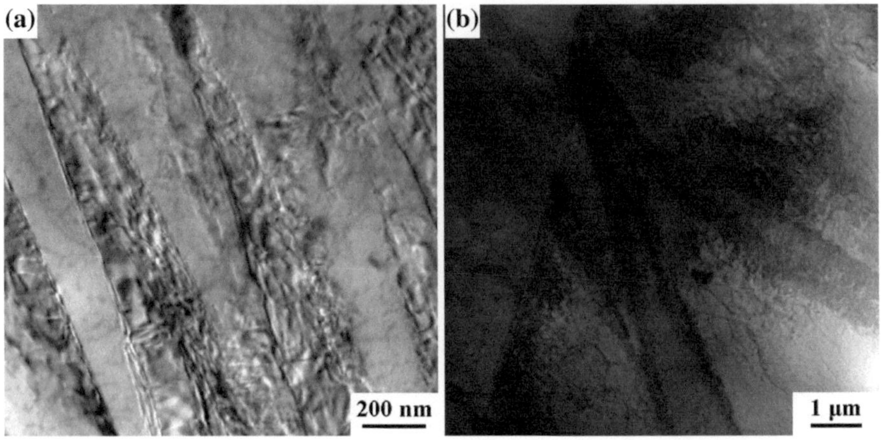

Fig. 4.9 TEM characterization at **a** UZ and **b** LZ of the joint subjected to FSP sample, showing the different sizes of twin structures [15]

yield strength and the tensile strength of the joint were improved. The authors concluded that the use of additional friction stir processing on a friction stir welded joint proved to be a feasible and an effective technique to enhancing the joint performance using the optimized process parameters compared to other joint improving techniques. Figure 4.9 showed a TEM image of the different sizes of twins in the upper zone (UZ) and the lower zone (LZ). The twin boundaries can effectively hinder the movement of dislocations and hence improved strength.

Balakrishnan et al. [16] also used friction stir processing as a secondary processing technique to improve the properties of cast aluminium matrix composites (AMCs) of Al/(0e15 wt%) Al3Fe AMCs. The as-cast composites are characterized with coarse grains, pores, segregation and sharp-edged particles. After the friction stir processing, the distribution of the particles became homogeneous as against the inhomogeneous particle distribution in the cast composite. Also, the FSP helped to remove the casting defects such as pores and make the sharp edges of the Al_3Fe particles into a near-spherical shape due to severe plastic deformation during the FSP. It reduced the sizes of grains of the cast composite into extremely fine grains because of the severe plastic deformation and dynamic recrystallization. This resulted in microstructural changes leading to tensile strength and ductility improvement. The fracture surfaces of the as cast composite and FSP samples as shown in Fig. 4.10 showed that there is an improvement in ductility.

Babu et al. [17] investigated the influence of travelling speed and rotation speed on properties of friction stir processed AZ31 magnesium alloys. The authors studied three different travelling speeds of 20, 30 and 40 mm/min; and two different rotation speeds of 600 and 800 rpm. The results showed that friction stir processing produced grain refinement and the smallest size grain is achieved at high rotation speed and high feed rate or travel speed resulting superior mechanical properties that helped to

Fig. 4.10 SEM micrographs of fracture morphology of: **a** cast AA6061, **b** friction stir processed AA6061, **c** cast AA6061/15 wt% Al_3Fe and **d** friction stir processed AA6061/15wt% Al_3Fe [16]

facilitate superplastic forming. The SEM images of the sizes of grains at different process parameters are shown in Fig. 4.11.

Qin et al. [18] studied the corrosion property of Friction stir processed magnesium alloy and nano-hydroxyapatite (nHA) particles through a low and high rotation speed of 300 and 600 rpm. The results showed that there is grain refinement with the distributed nanoparticles that resulted in improved hardness and corrosion resistance. The high corrosion resistance of FSPed samples was attributed to the grain refinement and the low surface energy. Multiple passes of FSP can further produce grain refinement, which can further improve the properties of the processed samples. The influence of process parameters and number of passes on the microstructure are shown in Fig. 4.12.

The use of friction stir processing for crack repair in stainless steel reactor has been reported by Miles et al. [19]. Repairing cracks in nuclear reactor components is often a challenge due to the large heat-affected zone that is created when such repair is carried out using the traditional fusion welding technique resulting in the development of intergranular cracks in the heat-affected zone (HAZ). This problem can be offset by using friction stir processing that operates at much lower peak temperatures than fusion welding. Crack repair of irradiated 304L stainless steel using the friction stir processing was investigated by Miles et al. [19] and developed a numerical simulation model temperature prediction and recrystallized grain size

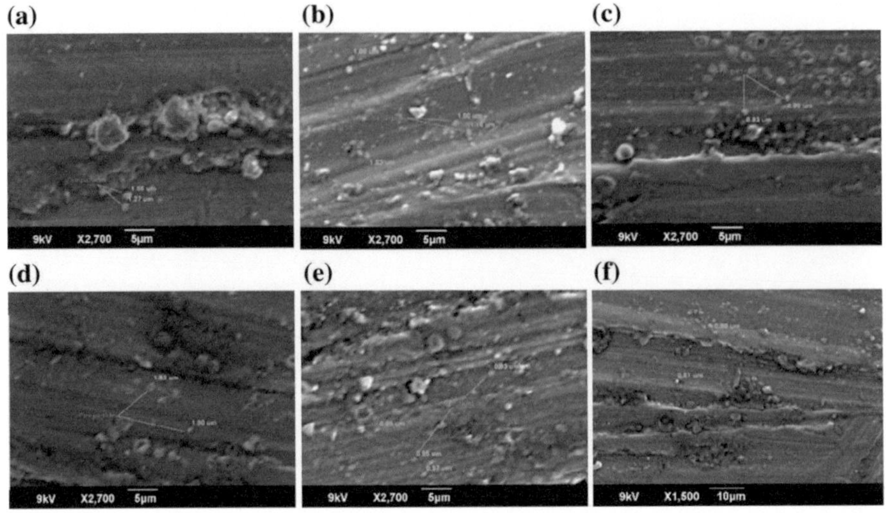

Fig. 4.11 SEM images representing smallest precipitation sizes obtained from friction stir processing of AZ31 Mg alloy processed at different spindle speeds and feed rates **a** 600 rpm 20 m/min **b** 600 rpm 30 m/min **c** 600 rpm 40 m/min **d** 800 rpm 20 m/min **e** 800 rpm 30 m/min **f** 800 rpm 40 m/min [17]

in the stir zone. The results showed that the friction stir processing can be used to repair cracks in nuclear reactor, The experimental results was found to be in close agreement with the predicted values by the developed numerical model.

Zhang et al. [20] fabricated a carbon nanotubes reinforced aluminium matrix nanocomposite using friction stir processing. The influence of energy input on microstructures and mechanical properties during friction stir processing of nanocomposites was studied. The results showed that the higher the input energy, the coarser the grains. Although the carbon nanotubes incorporated into the aluminium matrix by friction stir processing were evenly distributed at higher energy input which enhanced the mechanical properties of the nanocomposites fabricated. Figure 4.13 shows the fracture surfaces of the 3.2vol.%CNTs/Al nanocomposites fabricated with different energy inputs showing ductile fracture as a result of high elongation of the three nanocomposites.

The influence of Carbon nanotubes (CNTs) and or micron-sized silicon carbide (SiC) particles reinforced with Al5083 matrix to develop mono- and hybrid composites using the friction stir processing on the evolving microstructure and mechanical properties was investigated by Jain et al. [21]. The results revealed that significant grain refinement, grain fragmentation, as well as re-precipitation of second phase particles (see Fig. 4.14) were produced in the processed composite leading to improved mechanical properties.

Li et al. [22] studied the mechanical and cavitation erosion behaviour of friction stir cast nickel aluminium bronze (NAB). The study revealed that the microstructure of friction stir processed sample as homogeneous and refined microstructure with

(a)
(b)
(c)
(d)
(e)

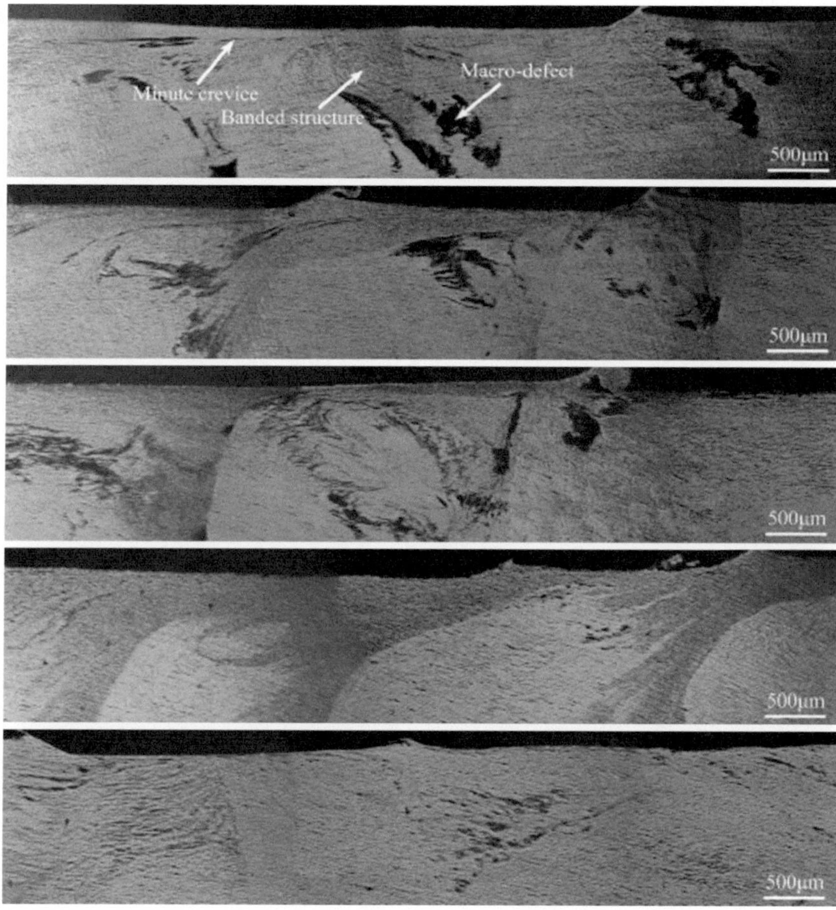

Fig. 4.12 Macrographs of cross section in FSPed ZK60-nHA samples at **a** 3000 rpm–100 mm/min, 1 pass, **b** 4000 rpm–100 mm/min, 1 pass, **c** 5000 rpm–100 mm/min, 1 pass, **d** 6000 rpm–100 mm/min, 1 pass, **e** 6000 rpm–100 mm/min, 2 pass [18]

(a) (b) (c)

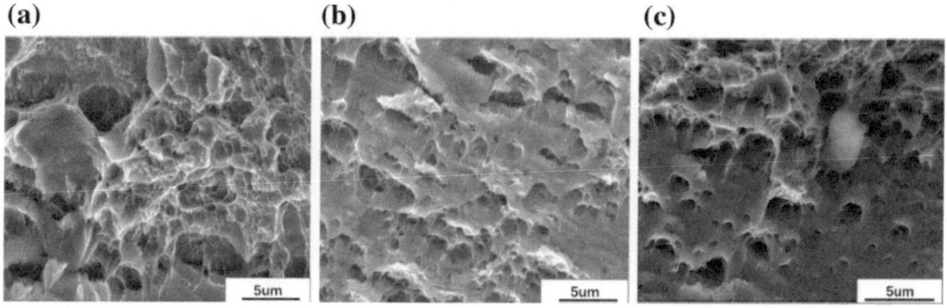

Fig. 4.13 Fracture surfaces of CNTs/Al nanocomposites. **a** 3.2%vol.% CNTs/Al-L. **b** 3.2%vol.% CNTs/Al-M. **c** 3.2%vol.% CNTs/Al-H [20]

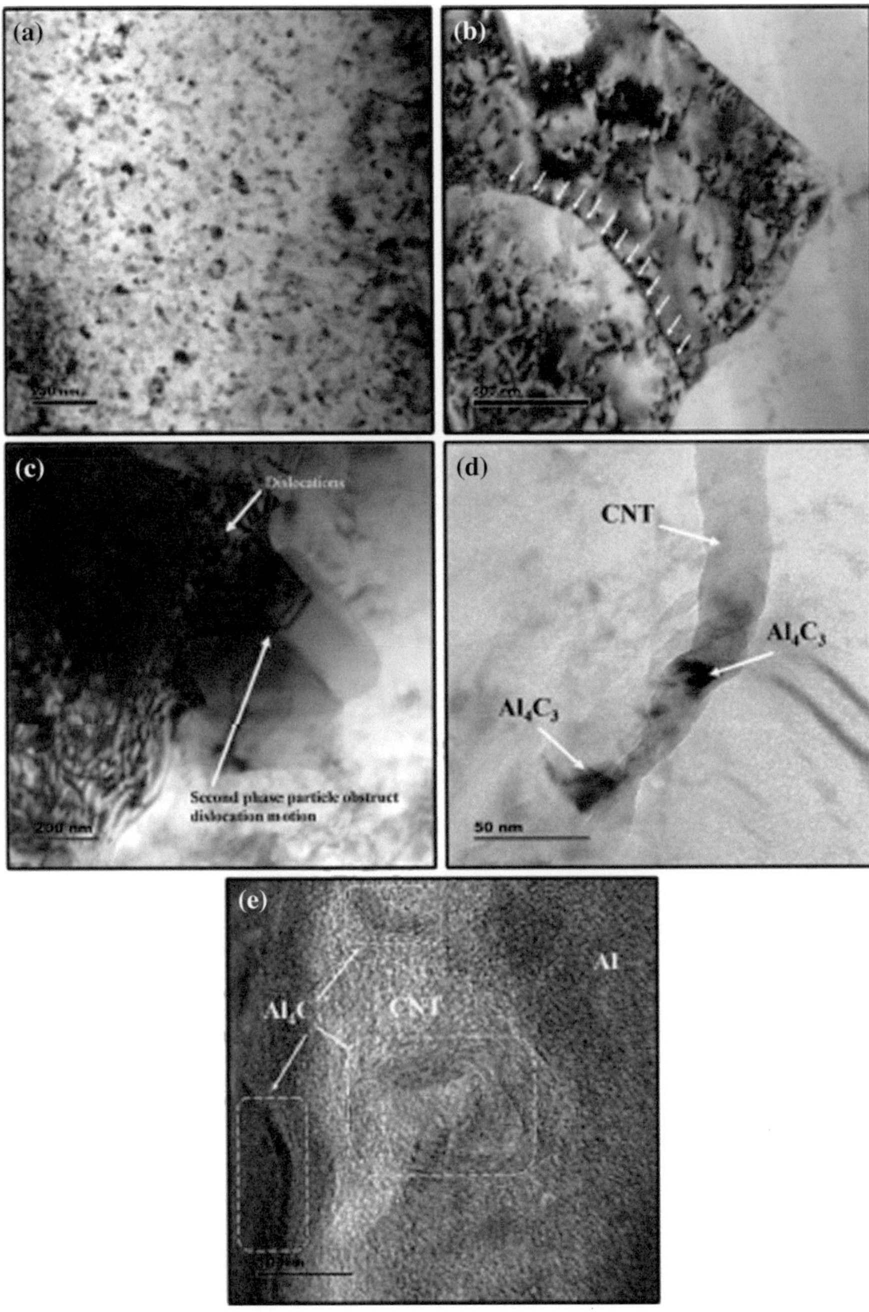

Fig. 4.14 Bright field TEM images of Al5083-SiC/CNTs showing **a** Fine second phase particles in stir zone, **b** dislocation rearrangement into subgrain boundaries (white arrows), **c** second phase fine particle obstruct dislocation motion, **d** survived twisted CNTs with Al4C3 phase, **e** interface between CNTs and Al matrix leading to Al4C3 phase formation [21]

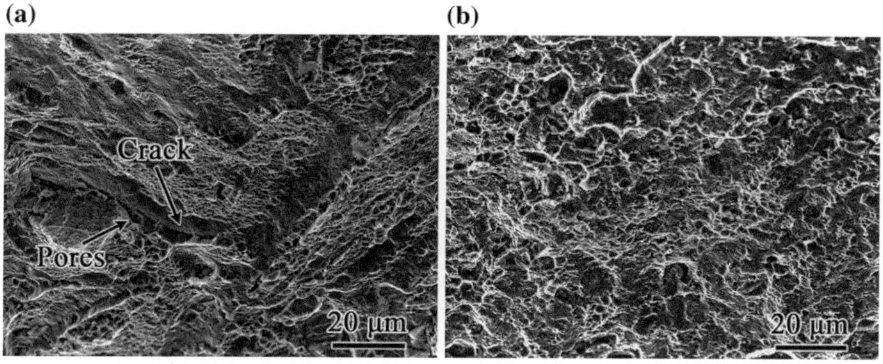

Fig. 4.15 SEM images showing the fracture surface morphologies for the **a** as-cast NAB and **b** FSP NAB [22]

improved microhardness, yield strength, ultimate tensile strength, and elongation when compared to that of the as-cast NAB alloy. The cavitation erosion resistance property of the friction stir processed samples are also found to be improved as compared to the as-cast NAB alloy. The result of tensile test fracture surfaces of the friction stir processed sample and as-cast sample are shown in Fig. 4.15. The large cracks and many dimples seen on as-cast sample (see Fig. 4.15a) is an indication of less plastic deformation before failure and hence lower tensile strength. The larger number of large dimples seen on the fracture surface of the friction processed sample as shown in Fig. 4.15b is an indication of better tensile property. The fracture surfaces are in good agreement with the tensile test results.

Peng et al. [23] also investigated microstructure and mechanical properties of friction stir processed AZ31 magnesium alloy. The results showed an improved microhardness in the stir zone with the top layer showing the highest microhardness value due to the high dislocation density at this layer. The tensile property was also found to be improved significantly due to the microstructure and texture modification. The yield strength of the stir zone is lower as a result of weak texture in the stir zone, and a higher elongation was observed as a result of the easier activation for basal slip. Akinlabi et al. [24] investigated the influence of processing parameters on the wear resistance behaviour of friction stir processed Al–TiC composites. The wear resistance property was found to be improved as a result of the TiC powder addition. The readers can refer Ebrahimi and Par [25] and Weglowski [26] on review of some research works in friction stir processing for further reading.

Hangai et al. [27] studied the fabrication of functionally graded aluminium foam consisting of A1050 and A6061 aluminium alloy using friction stir processing. The authors fabricated the functionally graded material using precursor and the bonding of the A1050 precursor to the A6061 precursor using the friction stir processing technique. The results showed that the Mg content gradually changed in the bonding region into Al showing a seamless functionally graded aluminium foam. Shi et al. [28] investigated the flow behaviour in friction stir processing using SiC particle as a tracer material during the process. The influence of the tool's rotation direction and initial position of the SiC powder on the distribution of the reinforcing particles was studied. The investigation revealed that the rotation direction of friction stir processing tool has a great influence on the dispersion and distribution of SiC particles. Counterclockwise rotation of the tool produced a more uniform distribution of the SiC particles when compared to the effect produced by the clockwise tool rotation. Also, the initial position of the reinforcing particles is also found to influence the distribution of the reinforcement particles. The reinforcing particles were found to found in a smaller area when the reinforcing particles are placed near the top surface while the distribution covered a much larger area when the reinforced particles are placed in the middle layer. A much wider area was covered with uniform distribution of the reinforcing materials when the particles are put in the middle layer of the plate with the tool rotation direction in counterclockwise direction. The study concluded that by placing the reinforcing particles in the optimum position and by operating the friction stir tool in the counterclockwise direction, metal matrix composite can be produced with the reinforcing particles uniformly dispersed in a relatively large area and with no defect and with excellent tensile property using the friction stir welding technique.

Thomson et al. [29] studied the influence of processing parameters on the grain refinement of friction stir processing technique through modification of thick microstructure of a monolithic plate of a magnesium alloy. The results of this study revealed that using liquid nitrogen as the cooling medium in friction stir processed AZ31B–Mg alloy produced lower average grain sizes compared to those produced under ambient temperatures condition. Tool design was also found to have a significant influence on the average grain sizes that are produced during the friction stir processing due to a complex effect of strain and temperature generation by the tool geometry. The study concluded that to achieve ultra fine grains during friction stir processing, smaller tools with dimensions and lower tool rotation rate were proposed. Fatchurrohman and Abdullah [30] also studied the effect of process parameters, rotation speed, on tribology characterization of Al alloys Al-6061 and Al-6061 with 5 weight % Al_2O_3 produced using the friction stir processing. The parameters used in this study are the rotation speed 1000, 1200 and 1400 rpm while the traverse speed was kept at a constant value of 25 mm/min. The results showed a better surface finish with the friction stir processed samples than specimens with no reinforcement. The rotation speed was found to have a great influence on the coefficient of friction of the specimen. The higher the tool rotation speed, the lower the coefficient of friction of the specimen material. The study concluded that the friction stir processing can be used to locally modify the microstructures of materials in regions experiencing high frictional loading thereby significantly improving the overall performance of Al alloy matrix composite.

Ding et al. [31] also investigated the influence of friction stir processing on the phase transformation and microstructure of TiO_2 and Ti-6Al-4V alloy. The grain refinement in the stirring zone and the phase transformation in the matrix material were studied for biomedical applications using friction stir processing. Figure 4.16 shows the microstructure of Ti-6Al-4V sample after friction stir processing in the three zones. The TEM micrographs of three zones is shown in Fig. 4.17 showing the uniform distribution of the TiO_2 particles with an average size of 10 nm in SZ, crystallized structure, and dislocation wall around the interface. Figure 4.18a shows the microstructure of the stir zone consisting of the lamellar alpha plus beta ($\alpha + \beta$) phase seen in the primary β phase grains occurring due to high temperature in this region. The transition from β phase to α phase occurs at slow cooling rate producing the complete lamellar structure. Figure 4.18b shows the dislocation and stacking fault in a phase to balance the strain during phase transformation. The microstructure in the transition zone (TS) is shown in Figs. 4.18c and d while the microstructure in the base material is shown in Fig. 4.18e.

Datong et al. [32] investigated the superplastic behaviours of friction stir processing of magnesium alloys, Mg–9Al–1Zn, under water. The result showed that there was a significant grain refinement through the broken down of the primary and the second phases into small particles. The friction stir processes materials were found to exhibit excellent high strain rate superplasticity. A similar study was conducted by Asadi et al. [33]. The authors investigated the influence of number of friction stir passes and reinforcing particle types on the powder distribution pattern and properties of nano sized SiC and Al_2O_3 particles added to as-cast AZ91 magnesium alloy using friction stir processing. The result of this investigation revealed that ultra fine grains

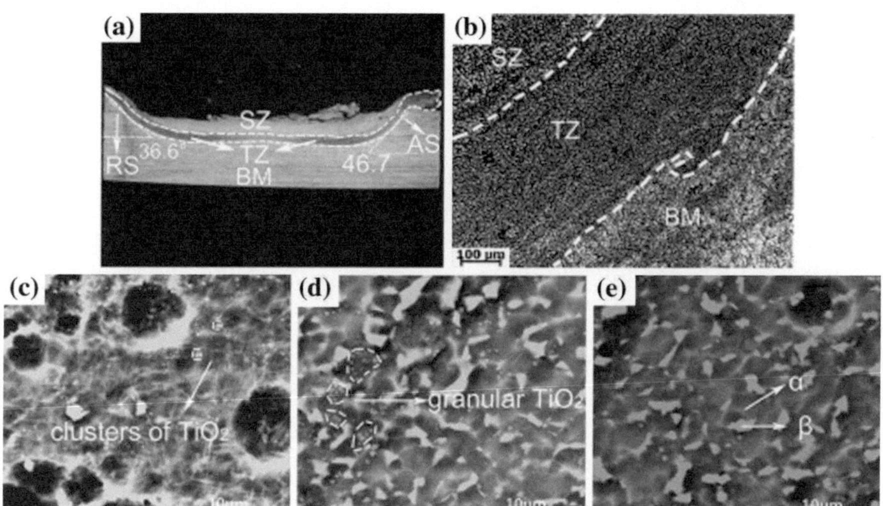

Fig. 4.16 Microstructure of different zones of the Ti-6Al-4V sample after friction stir processing: **a** macroscopic morphology, **b** overall microstructure of three zones, **c** stirring zone, **d** transition zone and **e** base material [31]

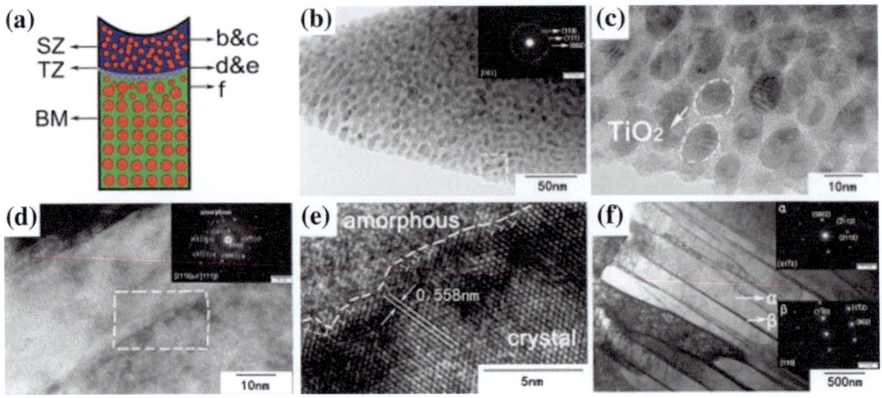

Fig. 4.17 TEM images of different zones of the Ti-6Al-4V sample after friction stir processing: **a** diagrammatic sketch of other figures, **b** nanocrystalline in SZ, **c** enlarged drawing for **b**, **d** interface between amorphous and crystal, **e** enlarged drawing for **d** and **f** structure between TZ and BM [31]

were achieved with higher hardness, higher strength and higher elongation in friction stir processed samples. Improved in particle distribution of reinforcing particles and finer gains with improved mechanical and wear resistance properties are achieved by increasing the number of passes in friction stir processing. The microstructure of sample in the stir zone with the SiC and Al_2O_3 particles reinforcements produced under two and three passes are shown in Figs. 4.19 and 4.20.

The distribution of reinforcement in the matrix become more uniformly dispersed as the number of passes in friction stir processing increases with improved properties. Samples produced with SiC reinforcement has higher strength and elongation than those of the samples produced using the Al_2O_3 particles as reinforcement. However, the wear resistance behaviour of SiC-reinforced samples reduces at higher number of passes in friction stir process.

Jiang et al. [34] studied the microstructure, mechanical and damping properties of a friction stir processed non-age-hardened Al alloy (5086) and an age-hardened Al alloy (7075). The result of this study showed that there is grain refinement after the friction stir processing with the grain size decreasing with decreasing tool rotation speed. At low tool rotation speed, the η phase precipitation in the 7075 alloy occurred thereby causing the micron-sized particles in the 5086 alloy to break up. Also, excellent mechanical and damping properties are obtained at low tool rotation speed of friction stir processed 5086 and 7075 alloys as a result of equilibrium in their grain boundaries, the fine grain, high fraction of high misorientation angle, low density of dislocations, and uniform particle distribution. Also, sample produced at low tool rotation speed exhibit higher internal friction value than those produced at higher tool rotation speed and that of the base material for the same alloy. Figure 4.21 shows the TEM micrograph of a 5086 base material, sample produced at tool rotation speed of 400 rpm and transverse speed of 100 mm/min (400-100-FSP 5086), sample produced at tool rotation speed of 1200 rpm and transverse speed

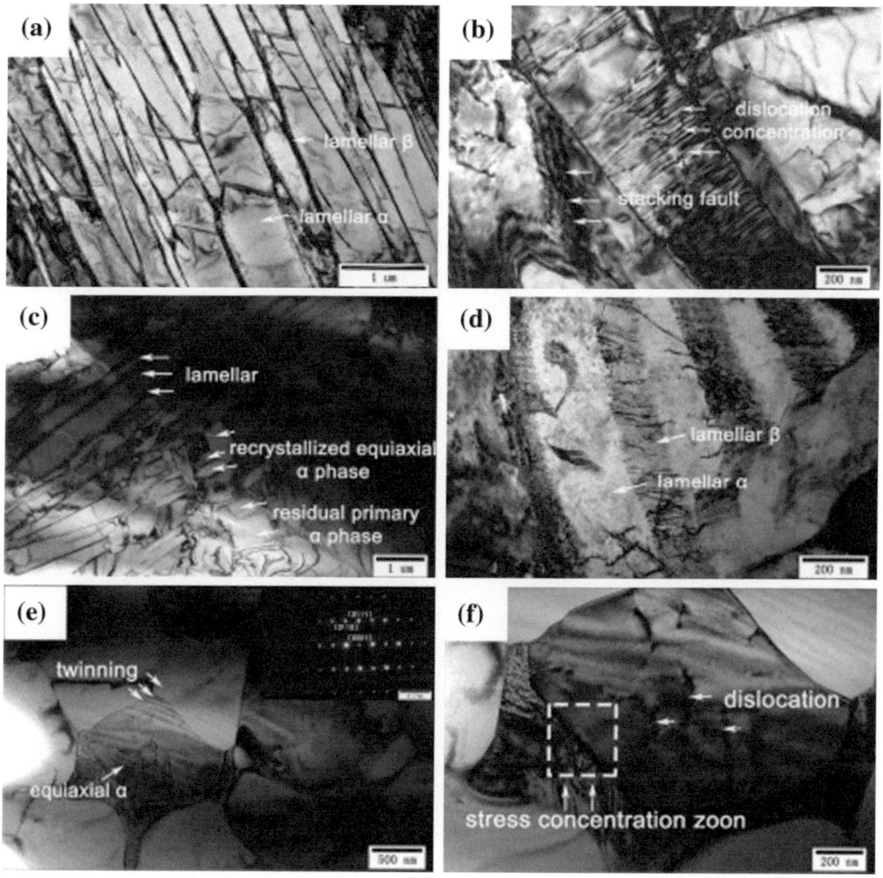

Fig. 4.18 **a** Lamellar a and b in SZ, **b** dislocation and stacking fault in SZ, **c** equiaxial and lamellar structure in TZ, **d** lamellar a and b in TZ, **e** equiaxial structure in BM and **f** dislocation and stress concentration zone in BM [31]

of 100 mm/min (1200-100-FSP 5086), and those of 7075 base material, sample produced at tool rotation speed of 400 rpm and transverse speed of 50 mm/min (400-50-FSP 7075), sample produced at tool rotation speed of 800 rpm and transverse speed of 50 mm/min (800-50-FSP 7075). Xu et al. [35] conducted a similar study investigating the properties of friction stir processed Mg-based metal matrix nanocomposites (MMNCs) of 6 vol.% SiC nanoparticles in Mg6Zn matrix using friction stir processing. The resulting friction stir processed Mg6Zn nanocomposites exhibit high hardness and high strength.

Naik et al. [36] investigated the influence of tool travel speed on the mechanical and wear resistance behaviour of friction stir processes Cu–0.18 wt%Zr alloy. The tool travel speed was varied between 50 and 200 mm/min while maintaining the

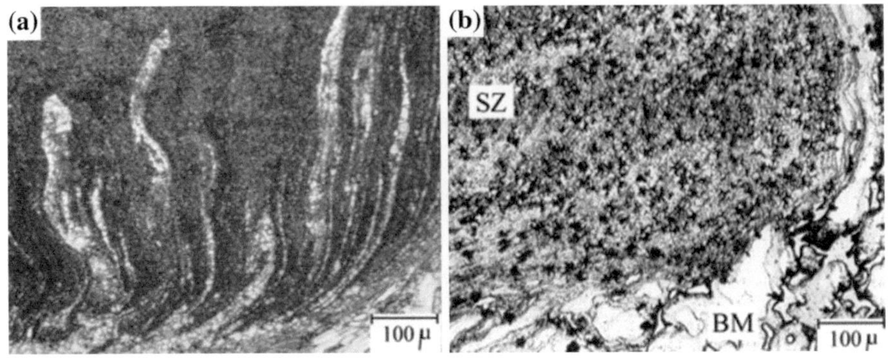

Fig. 4.19 Optical micrograph from the bottom corner of SZ at AS in the specimen produced by two FSP passes and **a** SiC and **b** Al_2O_3 particles [33]

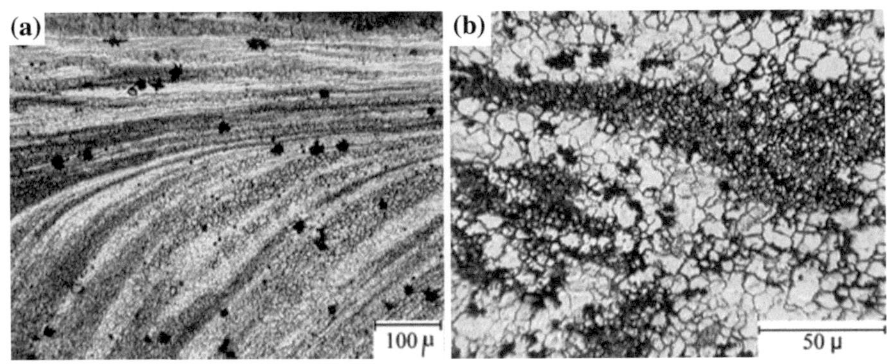

Fig. 4.20 Microstructures of SZ in the specimen produced by three FSP passes with Al_2O_3 particles **a** in low and **b** high magnification [33]

tool rotation speed at a constant value of 600 rpm. The results showed reduced grain sizes in the stir zone from 40.5 to 4.6 μm when the tool travel speed is increased. The hardness on the other hand was found to increase from initial 70–99 Hv with increasing tool travel speed. The coefficient of friction was decreased from 0.4 to 0.07 with increasing tool travel speed. The wear resistance was also found to be increased as the tool travel speed was increased. The microstructure of the base alloy and those of the friction stir processed samples at different tool travel speed are shown in Fig. 4.22.

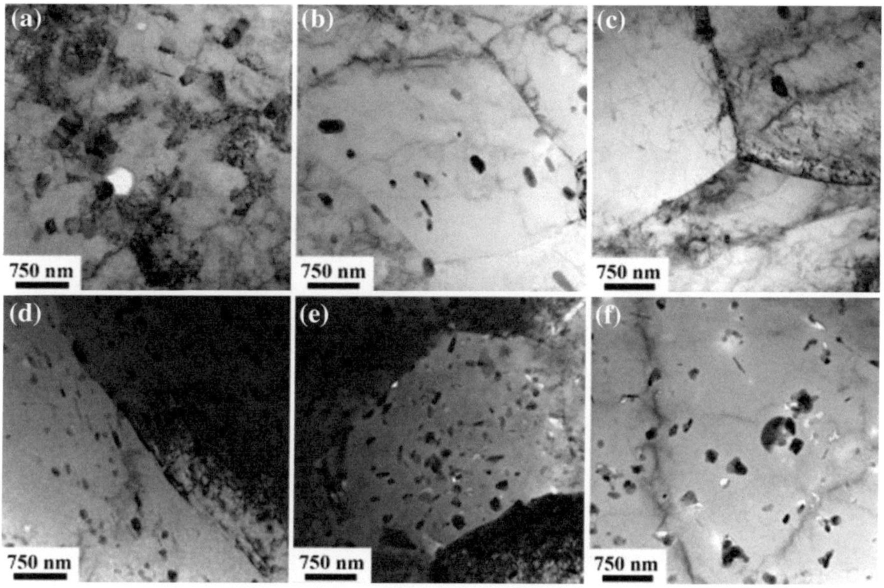

Fig. 4.21 **a** TEM micrographs of a 5086 BM, **b** 400-100-FSP 5086, **c** 1200-100-FSP 5086, **d** 7075 BM, **e** 400-50-FSP 7075, **f** 800-50-FSP 7075 samples [34]

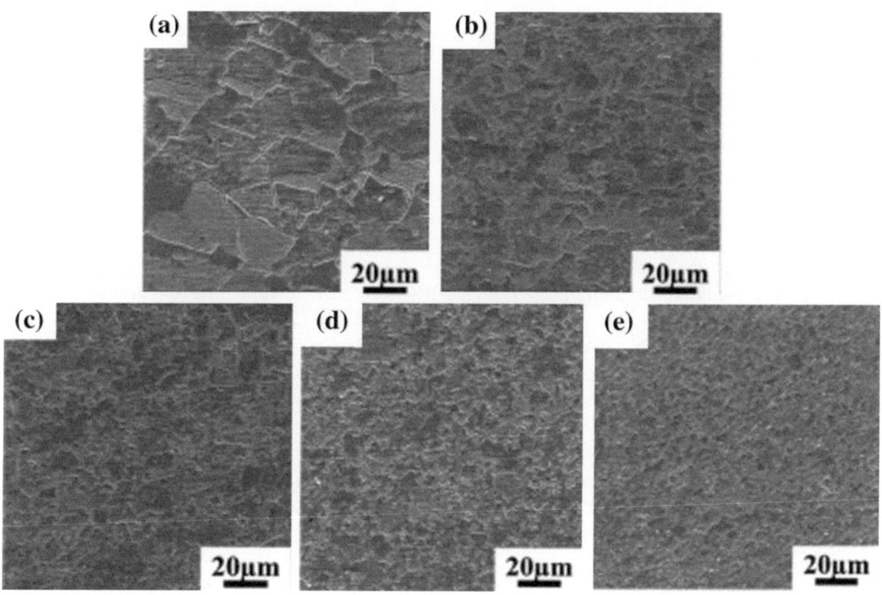

Fig. 4.22 Scanning electron micrographs of a base alloy and friction stir processed specimens with a travel speed of **b** 50, **c** 100, **d** 150 and **e** 200 mm/min [36]

4.6 Summary

The friction stir processing, an important grain refinement technique and is often used as secondary manufacturing process to improve properties of material. FSP is also used to improve the surface properties of material for production of composite surface with improved properties. The principle of operation of the FSP, advantages, limitation and areas of application are presented in this chapter. The importance of this surface modification technique has made it a research topic of choice in the research community. A number of research work has appeared in academic literature, some of these research works are also presented in this chapter.

References

1. R.S. Mishra, M.W. Mahoney, S.X. McFadden, N.A. Mara, A.K. Mukherjee, Scripta Mater. **42**, 163–168 (2000)
2. R.S. Mishra, M.W. Mahoney, Mater. Sci. Forum **357–359**, 507–512 (2001)
3. M. Dixit, J.W. Newkirk, R.S. Mishra, Scripta Mater. **56**, 541–544 (2007)
4. S.R. Sharma, Z.Y. Ma, R.S. Mishra, M.W. Mahoney, Scripta Mater. **51**, 237–241 (2004)
5. A.H. Feng, Z.Y. Ma, Scripta Mater. **56**, 397–400 (2007)
6. H. Izadi, A. Nolting, C. Munro, A.P. Gerlich, Effect of friction stir processing parameters on microstructure and mechanical properties of Al 5059, in *Trends in Welding Research, Proceedings of the 9th International Conference*, Chicago, Illinois, USA, pp. 382–386, 4–8 June 2012
7. H.H. El-Fahhar, H. Abd El-Hafez, M.M.Z. Ahmed, E. Ahmed, A. El-Nikhaily, The effect of tool geometry on the microstructure and mechanical properties of friction stir processed MIG-welded Aa5083-H111. IOSR J. Mech. Civil Eng. (IOSR-JMCE) **15**(4), Ver. II, 45–55 (2018)
8. S. Sinhmar, D.K. Dwivedi, Effect of weld thermal cycle on metallurgical and corrosion behavior of friction stir weld joint of AA2014 Aluminium alloy. J. Manuf. Process. **37**, 305–320 (2019)
9. N. Busu, S. Jaffarullah, C. Low, M. Saiful Bahari Shaari, Armansyah, A. Jaffar, A review of force control techniques in friction stir process. Procedia Comput. Sci. **76**, 528–533 (2015)
10. T. Wang, S. Shukla, M. Komarasamy, K. Liu, R.S. Mishra, Towards heterogeneous AlxCoCrFeNi high entropy alloy via friction stir processing. Mater. Lett. **236**, 472–475 (2019)
11. Yu. Chen, H. Wang, X. Wang, H. Ding, J. Zhao, F. Zhang, Z. Ren, Influence of tool pin eccentricity on microstructural evolution and mechanical properties of friction stir processed Al-5052 alloy. Mater. Sci. Eng., A **739**, 272–276 (2019)
12. W. Zhang, L. Tan, D. Nib, J. Chen, Y.-C. Zhao, L. Liu, C. Shuai, K. Yang, A. Atrens, M.-C. Zhao, Effect of grain refinement and crystallographic texture produced by friction stir processing on the biodegradation behavior of a Mg–Nd–Zn alloy. J. Mater. Sci. Technol. **35**, 777–783 (2019)
13. X. Jiang, N. Overman, N. Canfield, K. Ross, Friction stir processing of dual certified 304/304L austenitic stainless steel for improved cavitation erosion resistance. Appl. Surf. Sci. **471**, 387–393 (2019)
14. H.A. Deore, J. Mishra, A.G. Rao et al., Effect of filler material and post process ageing treatment on microstructure, mechanical properties and wear behaviour of friction stir processed AA 7075 surface composites. Surf. Coat. Technol. (2019). https://doi.org/10.1016/j.surfcoat.2019.05.048
15. Q. Shang, D.R. Ni, P. Xue, B.L. Xiao, K.S. Wang, Z.Y. Ma, An approach to enhancement of Mg alloy joint performance by additional pass of friction stir processing. J. Mater. Process. Technol. **264**, 336–345 (2019)

16. M. Balakrishnan, I. Dinaharan, R. Palanivel, R. Sathiskumar, Effect of friction stir processing on microstructure and tensile behavior of AA6061/Al_3Fe cast aluminum matrix composites. J. Alloys Comp. **785**, 531–541 (2019)
17. J. Babua, M. Anjaiah, A. Mathew, Experimental studies on friction stir processing of AZ31 Magnesium alloy. Mater. Today: Proc. **5**, 4515–4522 (2018)
18. D. Qin, H. Shen, Z. Shen, H. Chen, F. Li, Manufacture of biodegradable magnesium alloy by high speed friction stir processing. J. Manuf. Process. **36**, 22–32 (2018)
19. M.P. Miles, T.W. Nelson, C. Gunter, F.C. Liu, L. Fourment, T. Mathis, Predicting recrystallized grain size in friction stir processed 304L stainless steel. J. Mater. Sci. Technol. **35**, 491–498 (2019)
20. S. Zhang, G. Chen, J. Wei, Y. Liu, R. Xie, Q. Liu, S. Zeng, G. Zhang, Q. Shi, Effects of energy input during friction stir processing on microstructures and mechanical properties of aluminum/carbon nanotubes nanocomposites. J. Alloys Comp. (2019). https://doi.org/10.1016/j.jallcom.2019.05.269
21. V.K.S. Jain, K.U. Yazar, S. Muthukumaran, Development and characterization of Al5083-CNTs/SiC composites via friction stir processing. J. Alloy. Comp. (2019). https://doi.org/10.1016/j.jallcom.2019.05.232
22. Y. Li, Y. Lian, Y. Sun, Cavitation erosion behavior of friction stir processed nickel aluminum bronze. J. Alloys Comp. **795**, 233–240 (2019)
23. J. Peng, Z. Zhang, J. Huang, P. Guo, Y. Li, W. Zhou, Y. Wu, The effect of the inhomogeneous microstructure and texture on the mechanical properties of AZ31 Mg alloys processed by friction stir processing. J. Alloys Comp. **781**, 1074–1090 (2019)
24. E.T. Akinlabi, R.M. Mahamood, S.A. Akinlabi, E. Ogunmuyiwa, Processing parameters influence on wear resistance behaviour of friction stir processed Al-TiC composites, in *Advances in Materials Science and Engineering*, vol. 2014 (2014). http://www.hindawi.com/journals/amse/2014/724590/
25. M. Ebrahimi, M.A. Par, Twenty-year uninterrupted endeavor of friction stir processing by focusing on copper and its alloys. J. Alloys Comp. **781**, 1074–1090 (2019)
26. M.S. Węglowski, Friction stir processing—state of the art. Arch. Civil Mech. Eng. **18**, 114–129 (2018)
27. Y. Hangai, Y. Oba, S. Koyama, T. Utsunomiya, Fabrication of A1050-A6061 functionally graded aluminum foam by friction stir processing route. Metall. Mater. Trans. A **42**, 3585 (December 2011)
28. Q.-Y. Shi, K. Sun, W. Wang, G. Chen, Flow behavior of SiC particles as tracer material during the fabrication of MMCs by friction stir processing. Frict. Stir Weld. Process. 29–38 (2013)
29. B. Thompson, K. Dohertl, J. Su, R. Mishra, Nano-sized grain refinement using friction stir processing. Frict. Stir Weld. Process. **VII**, 9–19 (2013)
30. N. Fatchurrohman, A. Abdullah, Surface roughness and wear properties of Al–Al_2O_3 metal matrix composites fabricated using friction stir processing, in *Intelligent Manufacturing & Mechatronics*, Lecture Notes in Mechanical Engineering (2018), pp. 155–160
31. Z. Ding, C. Zhang, L. Xie, L.-C. Zhang, L. Wang, W. Lu, Effects of friction stir processing on the phase transformation and microstructure of TiO_2-compounded Ti-6Al-4V alloy. Metall. Mater. Trans. A **47**, 5675 (December 2016)
32. Z. Datong, C. Fang, L. Yuanyuan, High strain rate superplasticity of a fine-grained Az91 magnesium alloy prepared by friction stir processing, in *The 8th Pacific Rim International Congress on Advanced Materials and Processing* (2013), pp. 1065–1071
33. P. Asadi, G. Faraji, A. Masoumi, M.K. Besharati Givi, Experimental investigation of magnesium-base nanocomposite produced by friction stir processing: effects of particle types and number of friction stir processing passes. Metall. Mater. Trans. A **42**, 2820–2832 (September 2011)
34. H.-J. Jiang, B. Zhang, C.-Y. Liu, Z.-X. Yang, Z.-Y Ma, Mechanical and damping behavior of age-hardened and non-age-hardened al alloys after friction stir processing. Acta Metallurgica Sinica (English Letters) **32**, 1135–1141 (2019)

35. J. Xu, C. Cao, S. Das, L. Chen, C. Ma, R.S. Mishra, X. Li, High performance Mg6Zn nanocomposites fabricated through friction stir processing. Magnes. Technol. 383–386 (2015)
36. R. Bheekya Naik, K. Venkateswara Reddy, G. Madhusudhan Reddy, R. Arockia Kumar, Development of high-strength and high-electrical conductivity Cu–Zr alloy through friction stir processing. Trans. Indian Inst. Met. **72**(6), 1431–1435 (2019)

Chapter 5
Friction Stir Welding and Friction Stir Processing: Case Studies

Joining of aluminium to copper comes with a lot of combined benefits when compared to either of the materials. This chapter presents a case study on the analysis of aluminium and copper joints produced via friction stir welding technique. The welds were characterised through microstructure, tensile behaviour, X-ray diffraction and electrical resistivity. It was observed that a good mixing of both materials was achieved during the process. The tensile tests revealed that the samples were characterised with good joint efficiency and ductile behaviour when compared to the parent materials, the XRD revealed very low peaks of intermetallics compounds in the welds while the electrical resistivity results showed that there was an increase in the resistivity as the heat input increases and this output can be optimized.

5.1 Introduction

Friction stir welding technology is vital and a major breakthrough in the field of solid-state welding. Dr. Wayne Thomas and his team of The Welding Institute in Cambridge, United Kingdom invented the technology [1]. Since its invention in 1991 spanning about three decades, a number of researchers have shown interest and explored this technology and this is due to the huge benefits derived from the FSW technique. The technology is referred to as a green technology because no gases are evolved during the process and there are no toxic fumes or smoke produced during or after the welding process. The process is energy efficient and environmentally friendly compared to other conventional fusion welding methods. The technology has found applications in many industries including shipbuilding, tank containers, aerospace, trains and monorails. A typical example is the welding of panels of underground monorails referred to as Tube in the city of London, UK. The research and development studies commenced with joining similar materials [2–6] and then quickly advanced into the joining of dissimilar materials [7–9]. Many

E. T. Akinlabi and R. M. Mahamood, *Solid-State Welding: Friction and Friction Stir Welding Processes*, Mechanical Engineering Series,
https://doi.org/10.1007/978-3-030-37015-2_5

researchers with successful reports [10–15] have also conducted friction stir welding of aluminium and copper. There is a lot of potential application of Al and Cu, in particular, in electronic components and power generation industries. This is due to their lightweight and the superior electric and thermal conductivity of copper. One of the potential applications of dissimilar Al and Cu welds is the electrical connections in battery assemblies and in bus bars. However, a low weld strength can occur due to high metallurgical reactivity and affinity between Al and Cu, and this is due to the formation of hard and brittle IMCs at the interface resulting from heat input into the welds. It is also a challenge to attain friction stir welds of Al and Cu that are free from defects such as large void, cracks and other defects that are commonly present in the welds. Against this background, this chapter presents a case study of joining aluminium to copper using FSW process.

5.2 Experimental Methodology

The parent materials used in this research study are 3 mm × 100 mm × 600 mm aluminium alloy 5754 and pure copper. The copper material was placed at the advancing side while the aluminium alloy was placed at the retreating side. The tool pin was placed in the aluminium alloy at the joint interface and made to touch copper. The I-STIR Process Development System (PDS) friction stir welding machine as shown in Fig. 5.1 at the Nelson Mandela University, Port Elizabeth, South Africa was employed to perform the welds. The sheets were clamped firmly to the backing plate of the weld bed as shown in Fig. 5.2. The tool holder (see Fig. 5.3a) with a fixed pin and tool (see Fig. 5.3b) were used to perform the welds with an optimized tool tilt

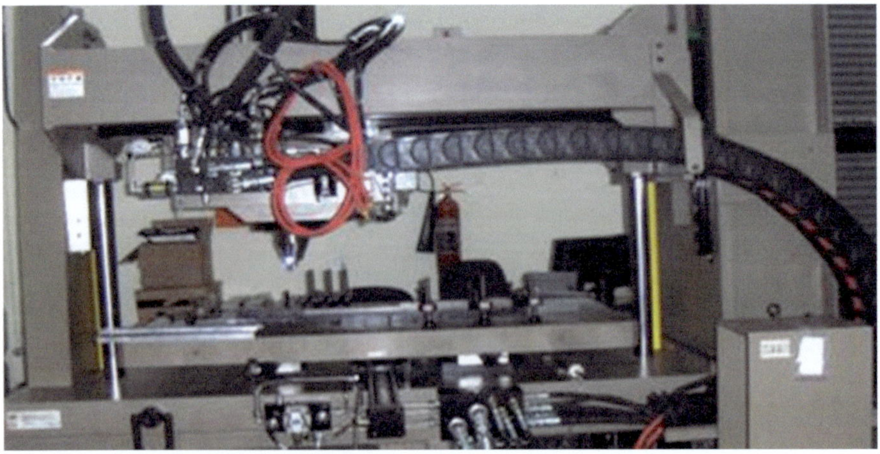

Fig. 5.1 Friction stir welding machine

Fig. 5.2 Parent materials clamp to the weld bed

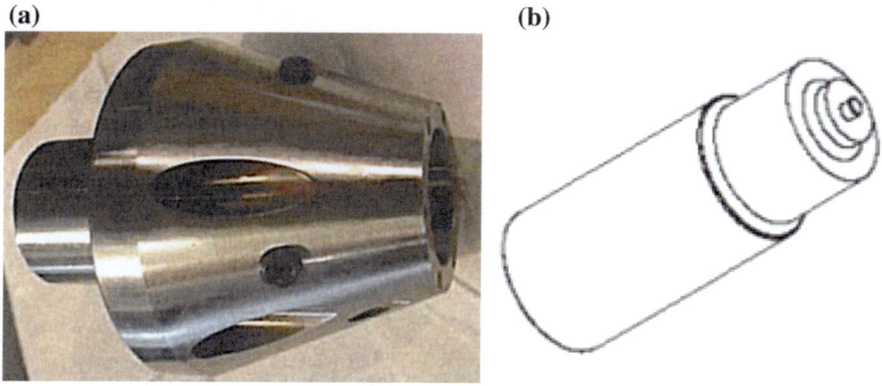

Fig. 5.3 **a** Tool holder and the **b** tool

angle of 2°. The features of the tool were threaded pin and concave shoulder. Three different shoulder diameters—15, 18 and 25 mm were used for the experiments.

The welds were produced by carefully selecting rotational speed and transverse speeds that depicted low, medium and high. The matrix is presented in Table 5.1.

The welds were characterised through the evolving microstructures by using modified Poulton's reagent to reveal the microstructure. Poulton's reagent consists of the following:

30 ml HCL, 40 ml HNO_3, 2.5 ml HF, 12 g CrO_3 and 42.5 ml of H_2O. The images were captured using the Olympus microscope and the images were analysed using

Table 5.1 Matrix of processing parameters

Rotational speed (rpm)	Transverse speed (mm/min)		
600	50	150	300
950	50	150	300
1200	50	150	300

Gwyddion Software. A servo-hydraulic Instron 8801 tensile testing machine was used to conduct the tests according to ASTM E8 [16]. Bruker D8 Advance X-ray diffractometer was used to analyse the welds. The four-point probe metre was used to measure the electrical resistivity of the welds. The results obtained are presented and discussed in the next section.

5.3 Results and Discussion

5.3.1 Microstructural Characterisation

The microstructures of the parent materials—aluminium is shown in Fig. 5.4 and that of Copper is shown in Fig. 5.5. It was observed that the grains in the aluminium are elongated indicating a cold-rolled condition, while those of copper were equiaxed, which can be attributed to the normalized or annealed state of the material.

Figure 5.6 shows a 2D typical micrograph of the joint interface of Al and Cu during friction stir welding. The weld was produced at a rotational speed of 950 rpm and a feed rate of 150 mm/min using the tool with 18 mm diameter shoulder.

It is, however, apparent from this figure that very little information can be obtained quantitatively. The darkish black colour is the aluminium while the reddish colour is the copper. The Gwyddion Software 2.51 version was, nevertheless useful in the quantification aspect of the micrograph. The flow pattern of the aluminium during

Fig. 5.4 Microstructure of 5754 AA (x500)

Fig. 5.5 Microstructure of copper (x500)

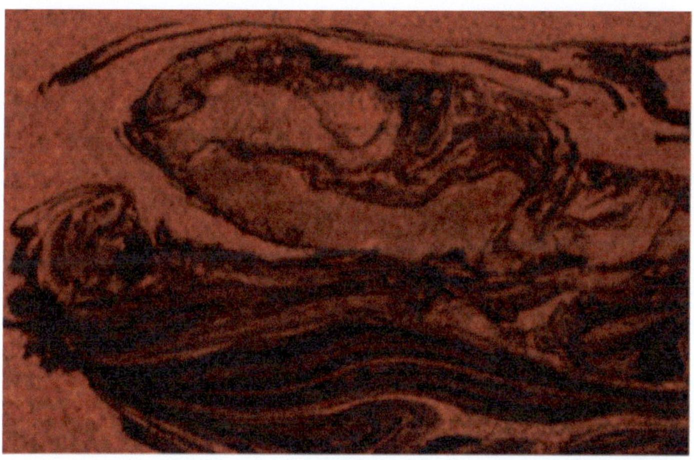

Fig. 5.6 2D Micrograph of friction stir welded Al and Cu produced at 950 rpm and 150 mm/min

the FSW process is due to a huge difference in the melting temperatures of Al and Cu with aluminium being a softer material when compared to copper. The 3D image of the weld is presented in Fig. 5.7 with aluminium showing as a darkish colour and flows in grooves of copper, which is a lighter phase as indicated.

Some distinguished statistical quantities were also obtained from the derived Gwyddion images to get value-related properties, area-related properties, boundary-related properties, volume-related properties, position-related properties and slope-related properties only to mention a few. This information is depicted in Fig. 5.8. This figure shows that the maximum peak height (Sp) of the matrix is 0.2332 μm

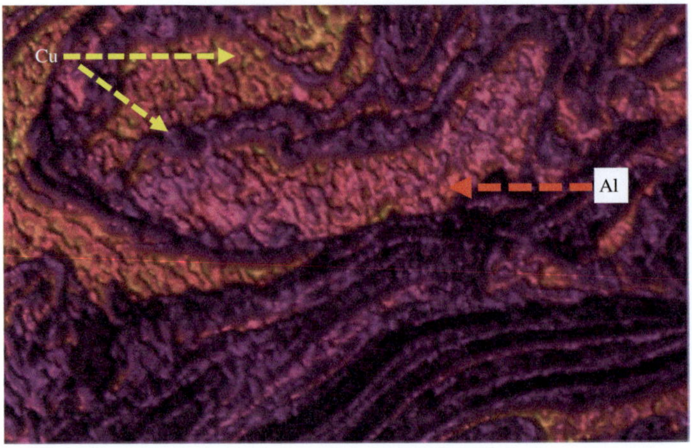

Fig. 5.7 3D image of friction stir weld aluminium and copper

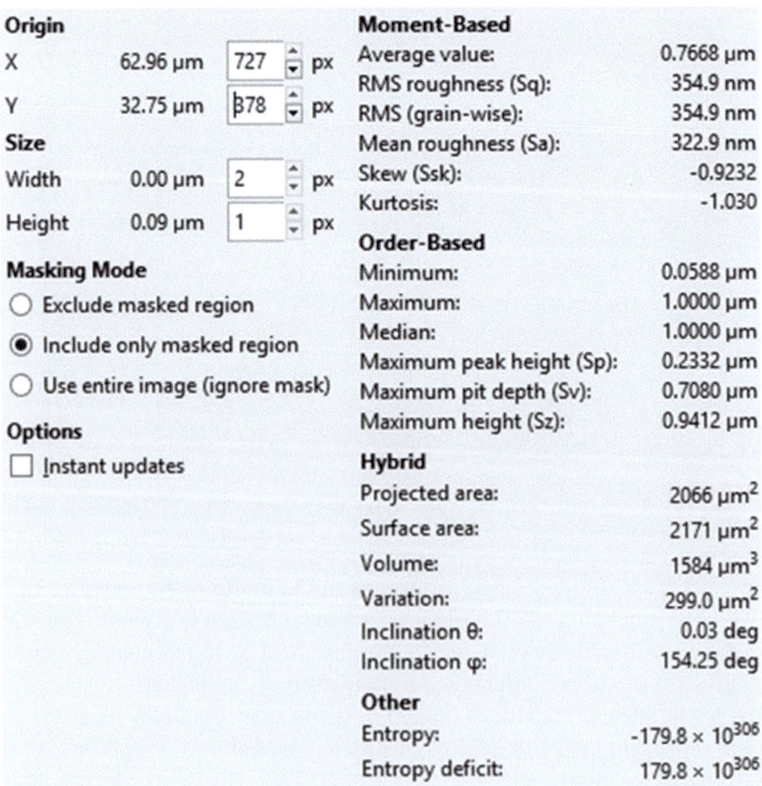

Fig. 5.8 Statistical quantities of the extracted masked region of stir zone during friction stir welding of aluminium and copper

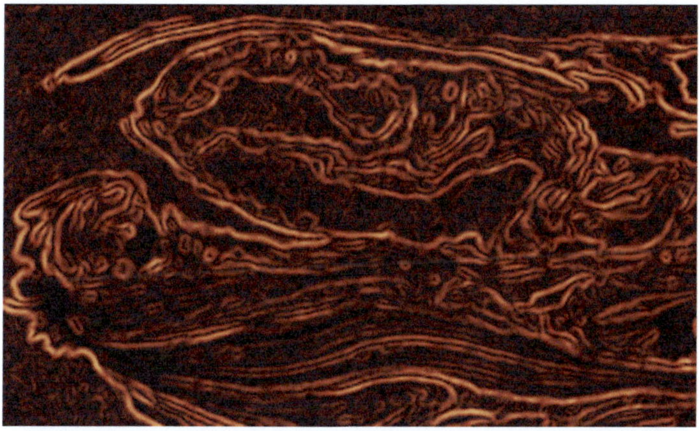

Fig. 5.9 Distortion effect of the grain structure during FSW of aluminium and copper

and the maximum height (Sz) is 0.9412 μm. The pit depth (Sv) is 0.7080 μm. The surface area of this masked region was 2066 μm^2 and the volume was 1584 μm^3.

Figure 5.9 shows an image of the distortion effect of the grain structure due to stirring and heat generation during FSW of aluminium and copper. The distortion effect in the grains is due to the effect of tool pin and shoulder during the FSW process. It is obvious from this figure that the grains were recrystallized during the FSW process.

5.3.2 X-ray Diffraction Analysis

The diffractogram of the weld produced with the 18 mm shoulder diameter tool at 950 rpm and 150 mm/min is presented in Fig. 5.10.

Apart from the main peaks of aluminium and copper, the additional diffraction peaks shown in the figure are due to the presence of intermetallic compounds such as Al_2Cu or Al_4Cu_9. These compounds were identified as far as possible by matching the data files. It is noted that the peaks were low and weak in this particular weld and this is almost the same for all the welds produced in this research study. The low peaks are due to low concentrations of the intermetallic compound present in the welds.

5.3.3 Tensile Behaviour

The tensile data of all the friction stir welds produced are presented in Figs. 5.11 and 5.12.

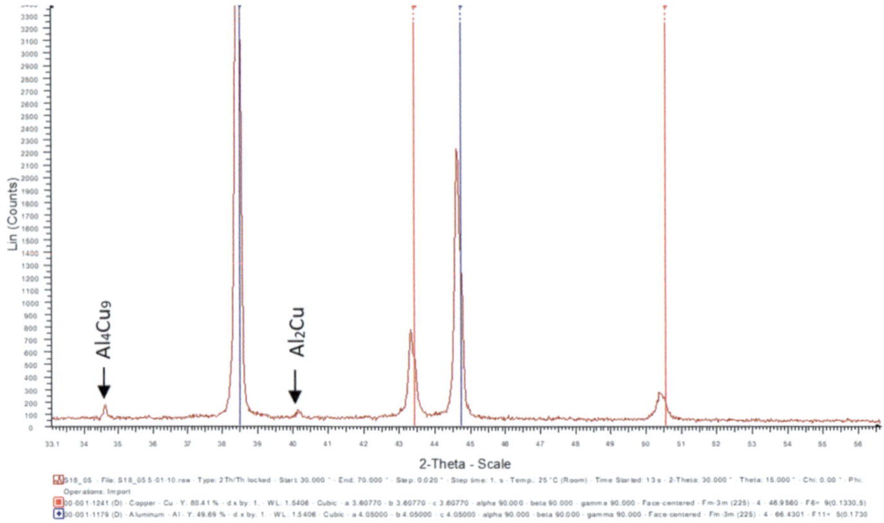

Fig. 5.10 Diffractogram of the weld produced with the 18 mm shoulder diameter tool at 950 rpm and 150 mm/min

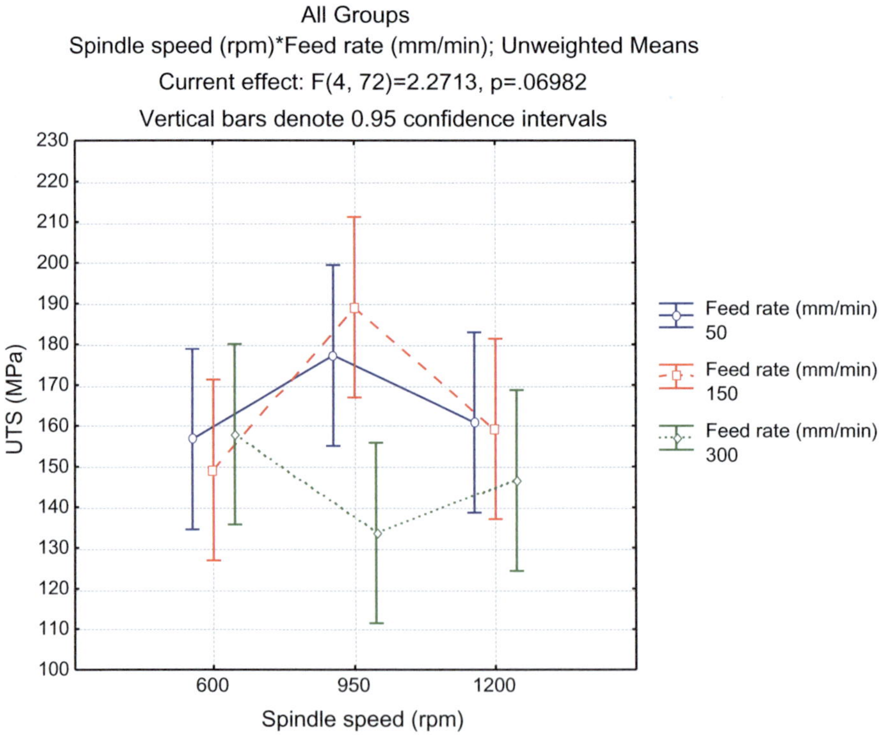

Fig. 5.11 Ultimate Tensile Strength (UTS) of all the FSwelds of Al and Cu produced

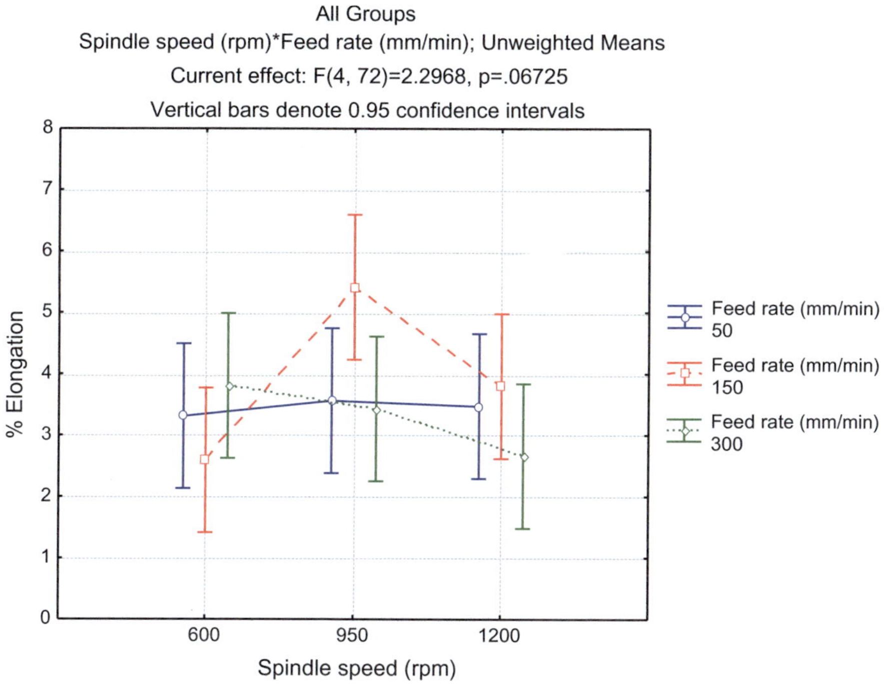

Fig. 5.12 Percentage elongation of all the FSWelds of Al and Cu produced

Trends apparent from the graphical representations of the ANOVA results of the tensile data as shown Fig. 5.11 and Fig. 5.12, respectively, is that the Ultimate Tensile Strength (UTS) and the percentage elongation of all the welds produced irrespective of the shoulder diameter employed (15, 18 and 25 mm) are higher at 950 rpm when compared to 600 and 1200 rpm, this is a reflection of the inherent joint integrity of the welds. This is an indication that the optimum process window for spindle speed to produce a good weld is at medium rotational speed of 950 rpm. In addition, it was observed that the UTS decreases as the feed rate increases from 50 to 300 mm/min.

5.3.4 Electrical Resistivity

The electrical resistivity data are presented in Fig. 5.13.

It was observed that the electrical resistivities of the welds increases as the heat input into the welds increases at a micro-ohm scale. The electrical resistivities of the welds ranged between 0.087 and 0.1 $\mu\Omega$ and no significant difference was found in the electrical resistivities measured from one weld to the other, this is expected due to the presence of the intermetallics compounds, which resists the flow of electricity.

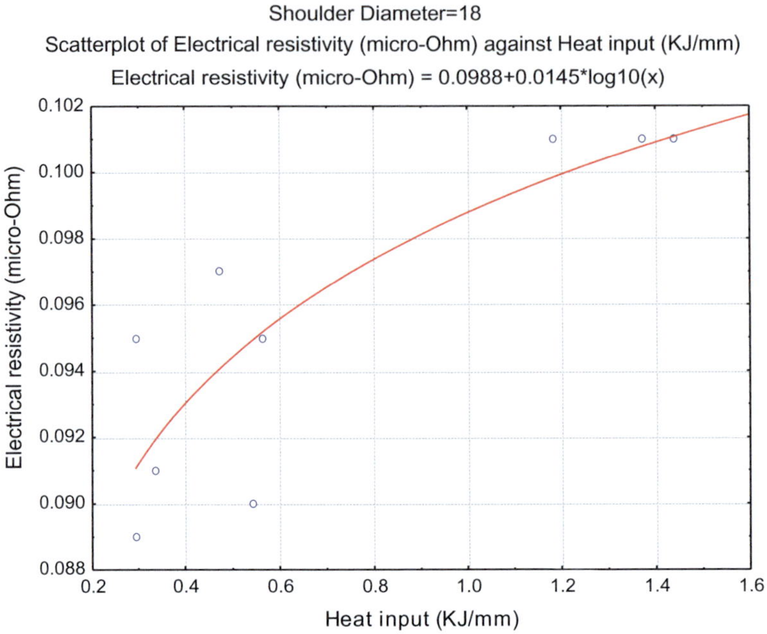

Fig. 5.13 Electrical resistivity data of FSWelds of Al and Cu

However, the low peaks of the intermetallics present can be correlated to the low resistivities measured in the welds.

5.4 Conclusions

The following conclusions can be drawn from the case study presented:

- That aluminium and copper welds produced via FSW has many potential benefits in industrial applications.
- The microstructural evolution of both materials joined revealed that there is a good mixing of both materials joined during the welding process.
- That the intermetallic compounds formed in the welds have very low peaks, hence the processing parameters considered in this study can be recommended.
- That the welds exhibited good tensile behaviour with sound joint integrities.
- The electrical resistivities of the welds are relatively low, although, this outcome can still be optimized.

References

1. W.M. Thomas, E.D. Nicholas, J.C. Needham, M.G. Murch, P. Templesmith, C.J. Dawes, Improvements relating to friction welding. International Patent Application, PCT/GB92/02203 (Patent), December 1991
2. V. Rajikumar, M. Venkatesh Kannan, A. Natarajan, Friction stir welding of aluminium alloys, in *Aluminium Alloys—Recent Trends in Processing, Characterisation, Mechanical Behaviour and Applications*, IntechOpen, 2017
3. V. Patel, W. Li, G. Wang, F. Wang, A. Vairis, P. Niu, Friction stir welding of dissimilar aluminium alloy combination: state-of-the-art, Metals. **9**(3), 270 (2019). https://doi.org/10.3390/met9030270
4. V. Malik, S.N.K.H. Suresh Hebbar, S.V. Kailas, Investigations on the effect of various tool pin profiles in friction stir welding using finite element simulations. Procedia Eng. **97** (2014). https://doi.org/10.1016/j.proeng.2014.12.384
5. N. Ravinder Reddy, G. Mohan Reddy, Friction stir welding of aluminium alloys—a review. Int. J. Mech. Eng. Technol. **7**(2), 73–80 (2016)
6. M. Arun Siddharth, V.V. Ramalingam, P. Ramasamy, Simulation of friction stir welding of aluminium alloy AA5052—tailor welded banks, in *Intelligent Systems and Applications* (Springer, 2020). ISBN 978-3-030-16660-1
7. K. Nagendra Kumar, P. Ravikanth Raju, Dissimilar materials of friction stir welding—overview. Int. J. Eng. Trends Technol. **44**(3) (February 2017)
8. O. Damola, E.T. Akinlabi, An overview on joining of aluminium and magnesium alloys using friction stir welding for automotive lightweight applications. Mater. Res. Express. (2019). https://doi.org/10.1088/2053-1591/ab3262
9. A. Mishr, Friction stir welding dissimilar metal: a review. Int. J. Res. Appl. Sci. Eng. Technol. **6**(1) (January 2018). ISSN 2321-9653
10. E.T. Akinlabi, S.A. Akinlabi, Friction stir welding of aluminium and copper: fracture surface characterizations. Presented at the *International Conference on Mechanical Engineering*, WCE, London, UK, 2–4 July 2014
11. D.S. Chaudhari, Joining of aluminium to copper by friction stir welding. Int. J. Innov. Res. Adv. Eng. (2014). ISSN 2349-2163
12. D. Aravinkumar, A. Balamurugan, A review on friction stir welding of dissimilar materials between aluminium alloys to copper. Int. J. Latest Res. Eng. Technol. **2**(2), 9–15 (February 2016). ISSN 2454-5031
13. S. Celik, R. Cakir, Effect of friction stir welding parameters on the mechanical properties of the Al–Cu butt joints. Met. Open Access Metall. J. **6**(6), 133 (2016)
14. N. Sharma, A.N. Siddiquee, Z.A. Khan, Friction stir welding of aluminium and copper—an overview. Trans. Non-Ferr. Metals Soc. China **27**(10)
15. E.T. Akinlabi, Effect of shoulder size on weld properties of dissimilar metal friction stir welds. J. Mater. Eng. Perform. **21**, 1514–1519 (2012)
16. Standard test methods for tension testing of metallic materials, E 8M-01ϵ1. Copyright ©ASM International, USA (2012)

Chapter 6
Friction Stir Processing Technology: A Case Study

In this chapter, a case study is presented of friction stir processing (FSP) employed to process pure aluminium—AA1050. The microstructural evolution was investigated using Optical Microscope (OM) as well as Scanning Electron Microscope (SEM). Tensile behaviours of the Friction Stir Processed (FSPed) AA1050 samples were examined and the fractography of the fracture surfaces was also studied. The results show via macrograph that FSPed AA1050 samples were defect free and the most refined zone was the heat-affected zone (HAZ) with an average grain size of 6.77 μm while the grain sizes of stir zone (SZ) and thermo-mechanically HAZ (THAZ) were 8.47 and 7.34 μm respectively. It was further established that the ultimate tensile strength (UTS) of the FSPed AA1050 samples was 88.04 MPa while that of the base metal was 85.04 MPa whereas the yield strength of the FSPed AA1050 was 77.3 MPa and that of the base metal was 73.1 MPa. It was noted that FSPed AA1050 has higher percentage elongation of 69.21%, whereas base metal has 67.08%. The overall results show that FSPed AA1050 has better tensile properties when compared to the unprocessed base metal as evident in the fractography. The fractured samples indicate that there was appreciable plastic deformation and necking before failure took place. The fractured analysis equally shows fibrous features with appreciable lip region at the outer boundary of the final fracture. These noticeable fibrous features are an indication that the crack propagation occurred slowly. The cracking pattern exhibited what is termed equiaxed dimples with prominent cup and cone features, a typical behaviour of ductile materials.

6.1 Introduction

Friction Stir Processing is a variant of Friction Stir Welding process, it is usually employed for surface modifications and reinforcements particularly in aluminium and its alloys. Aluminium metal matrix composites (Al-MMCs) are types of composite materials that have found vast application in automobile, construction,

E. T. Akinlabi and R. M. Mahamood, *Solid-State Welding: Friction and Friction Stir Welding Processes*, Mechanical Engineering Series,
https://doi.org/10.1007/978-3-030-37015-2_6

transportation, nuclear, aerospace, marine, aviation, defence, space and air vehicle, ballistic industries [1]. FSP is one of the essential classes of modern, advanced processing and manufacturing of engineering materials due to its remarkable integration of composites properties. Aluminium possess some outstanding properties such as excellent formability, low coefficient of thermal expansion, good tribological (wear and corrosion) integrity, high strength-to-weight ratio, high thermal and electrical conductivity, high fatigue strength, excellent damping capacities, high plastic flow strength, high specific stiffness and strength, and lots more which endeared its wide utilization in the aforementioned industries [2–4]. Lower series of aluminium alloys such as series 1xxx (pure aluminium), 2xxx (copper-based), 3xxx (manganese-based), 4xxx (silicon-based), 5xxx (magnesium) and 6xxx (silicon-based) have been reported to have poor corrosion resistance, low tensile strength, low thermal stability and low hardness among wrought alloys [5–7]. In contrast, high-grade aluminium alloys are said to have improved mechanical and chemical properties over the lower series such as 7xxx (zinc-based), 8xxx and 9xxx, this set of aluminium alloys have better and improved corrosion mitigation and high strength properties [8, 9]. It worth mentioning that lightweight materials offered an excellent and competitive advantage over conventional steel due to its low density and high specific strength [10, 11]. It has been foreseen that in the near future, steel might wholly be replaced with low-density and high-strength non-ferrous alloys materials like aluminium, magnesium as well as titanium alloys in the aerospace, defence, automobile and transportation sectors. In light of this, aluminium might be the one that will likely replace steel in these sectors because titanium is expensive while magnesium is self-explosive [12].

Pure aluminium in isolation lost the integrities and capacities in meeting the high demand for advancement in manufacturing and processing applications. Pure aluminium suffer poor tribological properties, stiffness, strength and lots more which made it unsuitable for most applications in its current form hence reinforcement is required to ameliorate this deficiency. In recent times, researchers had made debut findings into curbing and suppressing this menace via the application of reinforcement particles [13]. Addition of reinforcements into pure aluminium has led to the production of aluminium metal matrix composites (Al-MMCs). Reinforcement in Al-MMCs has overcome many deficiencies that engraved pure aluminium. Reinforcements provide Al-MMCs materials with a better strengthening at a low cost and with much better mechanical, metallurgical and electrochemical properties. It was revealed that application of reinforcements has led to high ductile properties in the Al-MMCs materials at the same time increased the modulus and the strength of the composites [14]. It was established by many researchers that reinforcements played critical roles in the modification of structural surface and texture of the reinforced metal matrix composite materials and how it enriched electrochemical, mechanical and metallurgical properties of the materials via intense, localized plastic deformation compared to the base material [15, 16]. Al-MMCs is the new set of composites material in non-ferrous metals group that proffer solutions to the problems inherent with the ductility of Aluminium. Several methods are used to produce MMCs aside friction stir processing, other methods include stir casting, powder metallurgy, squeeze casting as well as compo casting. All these methods had been exploited by researchers to fabricate MMCs reinforced with different kinds of organic and inorganic particles either in macro-, micro- or nanoparticles [17].

Friction stir processing (FSP) is an excellent surface modifier [18, 19] that was developed from friction stir welding (FSW) which plunged into the world in 1991 and was patented by the welding institute (TWI) of UK as a solid-state joining technique and this has received tremendous attention to fabricate metal matrix composites (MMCs) in recent years. In FSP, a groove (microchannel) that will enclose the second phase (reinforcement particles) may or may not be created on the substrate [20, 21] and a non-consumable processing tool with a define shoulder tool profile mostly cylindrical and an exclusively designed pin or probe profile which is rotated and plunged into base metal (BM) and travelled on the substrate (workpiece) surface in the processing direction. The processing tool has been known for two fundamental functions: (a) heating and (b) deformation of workpiece material [22]. The processing tool was designed for stirring and mixing of matrix base metal with the reinforcement materials which is the second phase to form matrix composite. The FSP is a thermo-mechanical process that involves the interactions (rubbing action) of the tool shoulder and the substrate which produces high heat as a result of friction between them leading to the softening of the material under the shoulder due to severe plastic deformation during tool rotating action and stirring process leading to the production of MMCs [23]. It has been established that FSP produces short route, homogeneity and refined microstructures as well as densification. Studies shown that mechanical and metallurgical properties of the processed zone can be controlled via the optimization of tool geometry such as shoulder diameter, probe length, probe profile, groove dimensions (width and depth) and processing parameters such as rotational speed, travel speed, plunge rate, heat input and cooling/heating methods [21, 23], the flowchart of the process parameters used in FSP is shown in Fig. 6.1. FSP has

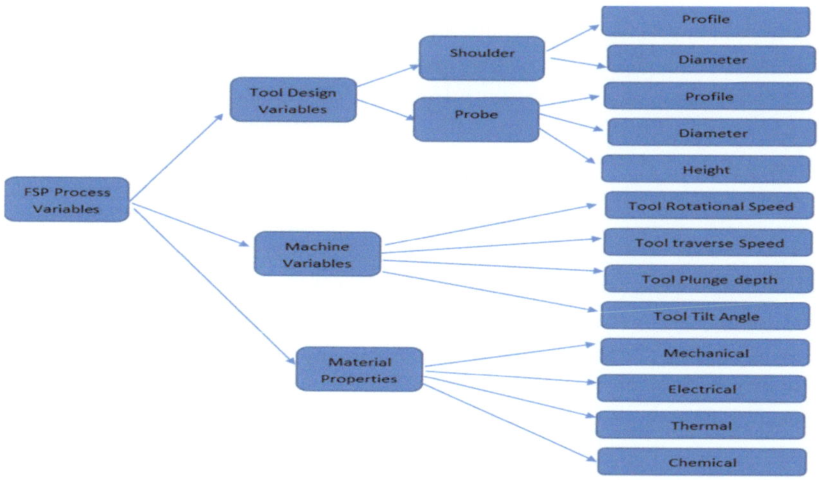

Fig. 6.1 Flowchart of process variables used in FSP

advantages over other manufacturing processes in a manner that minimizes defects and distortions in the material [24, 25]. FSP alters the physical properties of the base material without altering its physical state, these help engineers develop attributes such as 'high-state-rate superplasticity' [28]. The grain refinement occurs on the parent material which improves the properties of the first material while mixing with the second material (reinforcement). Subsequently, allows a variety of properties to be altered and this, in turn, improves its surface modification [29]. Different kinds of reinforcement have been employed since the debut entrance of FSP. It has been established that the most commonly applied reinforcing material during FSP is inorganic (metallic) powders, namely copper, graphite, iron, silicon carbide, nitrides, WC, titanium alloy, TiB_2, graphene, stainless steel, oxides such as SiO_2, Al_2O_3 [26, 30], etc. and some works had also been reported on organic powders (i.e. bioprocessing using agro-wastes powders) such as fly ash, palm kernel shell ash, coconut shell ash, rice husk ash, etc. Reinforcements can be grouped into two major classes and they are either ex situ or in situ. Ex situ is a technique of synthesis route by reinforcing the liquid or powdered metal, while the in situ process is a way of producing reinforcement compounds by reaction during processing such as using reactive gases [31].

It was established that the first article in which FSP first experiment was published in 1999 by Mishra, R.S [27, 32, 33] and since this appearance, thousands of articles, patents in hundreds, textbooks and many review papers on FSP had been in circulation. It is facile, update and cost-effective to prepare Al-MMCs using FSP [34]. A lot of benefits that engraved FSP include grain refinement, homogeneity of the processed zone, densification and homogenization of precipitates of aluminium alloys and composites materials [35]. FSP has been established for the improvement of the surface properties of the metal alloys, hardness, fatigue life, improvement in formability and ductility as well as increased strength without altering the bulk metal properties [22, 36]. This new and promising technology can be used to modify casting of alloy, increase the fatigue strength of the processed substrate, modify the chemical compositions of the surface layer, susceptibility to plastic deformation can be increased and also fabricating new alloys with special properties, as well as achieving the state of over-plasticity [37]. FSP has been used extensively for the modification of surface parts especially in aerospace, automation and railway for the fabrication of load-bearing components due to its environmental friendliness and high efficiency [38]. In the course of improving the surface quality by FSP, some challenges are encountered during the process of reinforcing the composite materials such as excessive wear during the fabrication of the composites as a result of abrasive effect of the reinforcing particles, tool wear, sticking of the substrate to the backing plate especially when the workpiece thickness less than 1 mm, challenges on how to improve fatigue property and joining strength, many optimizations may be required to obtain optimum parameters and this may lead to the usage and loss of many materials [33, 39].

6.2 Materials and Methods

6.2.1 Material

The parent materials used in this research was pure aluminium. The dimension of the test coupon for each plate was 250 × 210 × 6 mm and the length of the welds produced was 200 mm. The chemical composition of the parent material was confirmed, using a spectrometer and was found to conform to the standard pure aluminium specifications reported by Pereira et al. [40]. Table 6.1 shows the chemical composition of the parent material.

6.2.2 Friction Stir Processing Methodology

FSP of aluminium, alloy AA1050 was carried out on a 2 t linear numerical controlled (LNC) friction stir welding machine (FSW-M) which was manufactured by ETA Engineering Pvt. Ltd, Bangalore, India. and there is a load cell that is incorporated on FSW-M which is responsible for taking the forces along the Z-axis direction. The interface of the LNC-FSW-M is integrated with LabView software features capable for real-time data acquisition. The placement of pressure plate on the workpiece was to enable rigid clamping and at the same time proper and rigorous stirring for effective mixing of the materials through the tool pin penetration and the interaction of the shoulder with the workpiece via the tool translational and rotational [28, 41].

6.2.3 Process Parameters

In this present research work, the process parameters that were engaged are 1200 rpm of rotational speed, 50 mm/min of traverse speed, plunge depth of 0.3 mm with two passes having 100% overlapping on the previous pass and the distribution was carried in the following manner, first pass has 0.2 mm plunge depth while the second pass had 0.1 mm plunge depth, the tilt angle used was 3°. A cylindrical processing tool made of AISI H13 tool steel of shoulder diameter 18 mm, pin diameter 5 mm and a pin length of 5 mm with 10° taper was then employed for FSP.

Table 6.1 Chemical composition of Pure Aluminium—1050-F

Element	Si	Fe	Cu	Mn	Mg	V	Zn	Ti	Al
wt%	0.25	0.40	0.05	0.05	0.05	0.05	0.05	0.03	99.50

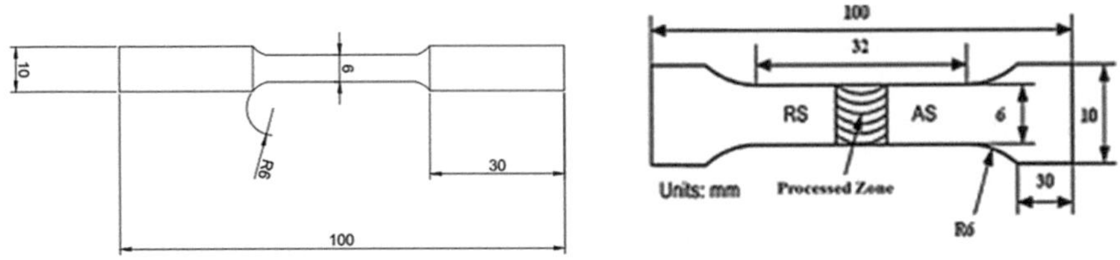

Fig. 6.2 Schematics of tensile samples (all dimensions in mm)

6.2.4 *Tensile Experiment*

Flat tensile samples of substandard size 100 mm long, 6 mm thickness were machined to size by the use of EDM and the tensile samples were experimented based on the ASTM E8M-13 standard specifications [42]. An Instron tensile testing machine was used to evaluate the longitudinal tensile properties. The FSPed materials were evaluated by testing three specimens in each condition to measure tensile strength, and yield strength was taken by constructing a straight line parallel to the initial linear portion of the stress–strain curve at 0.2% offset for each sample. The percentage of elongation of the FSPed samples were evaluated by measuring the final length of the failed specimens to determine the ductility of the samples. The elongation percentage is defined in the following Eq. (6.1):

$$\%El = \left[\frac{L_f - L_o}{L_o}\right] \times 100 \tag{6.1}$$

where L_o is the initial gage length, and L_f is the length of the gage section at fracture. Data showing a comparison of the base material and the FSPed samples in terms of tensile strengths, yield strength, and the elongation percentage. The schematic of the tested sample is shown in Fig. 6.2.

6.2.5 *Microscopy*

Optical microscopy was conducted using Olympus BX51M and Olympus SZX16 optical microscopes. The Olympus BX51M was used to observe the microstructures, while the Olympus SZX16 was employed to observe the sample macrographs. Digital output was captured and processed using Olympus Stream Essential software. Microstructures of the FSP samples were obtained for the BM, SZ, TMAZ and HAZ zones for comparison. The grain sizes of the different zones were carried out according to the standard test method for determining average grain size: ASTM E112-12 [43]. A TESCAN VEGA3 scanning electron microscope (SEM) was used to study and compare the microstructures observed in the base material. Vega TC software was used to acquire the image on the SEM.

6.3 Results and Discussions

In this research, the microstructural evolution of the base metal using OM and SEM were presented in Fig. 6.3a and b, respectively, while the macrograph of the FSPed AA1050 was presented in Fig. 6.3c. The average grain size for FSPed AA1050 at SZ, THAZ and HAZ was recorded in Table 6.2 while their corresponding microstructures were presented in Fig. 6.4. The procedures for microstructural evolution for FSPed AA1050 was illustrated in Fig. 6.5 while the force feedback generated during FSPed AA1050 was illustrated in Fig. 6.6. Tables 6.3, 6.4 and 6.5 represent the tensile properties results for the base metal —AA1050 and FSPed AA1050 and the summary

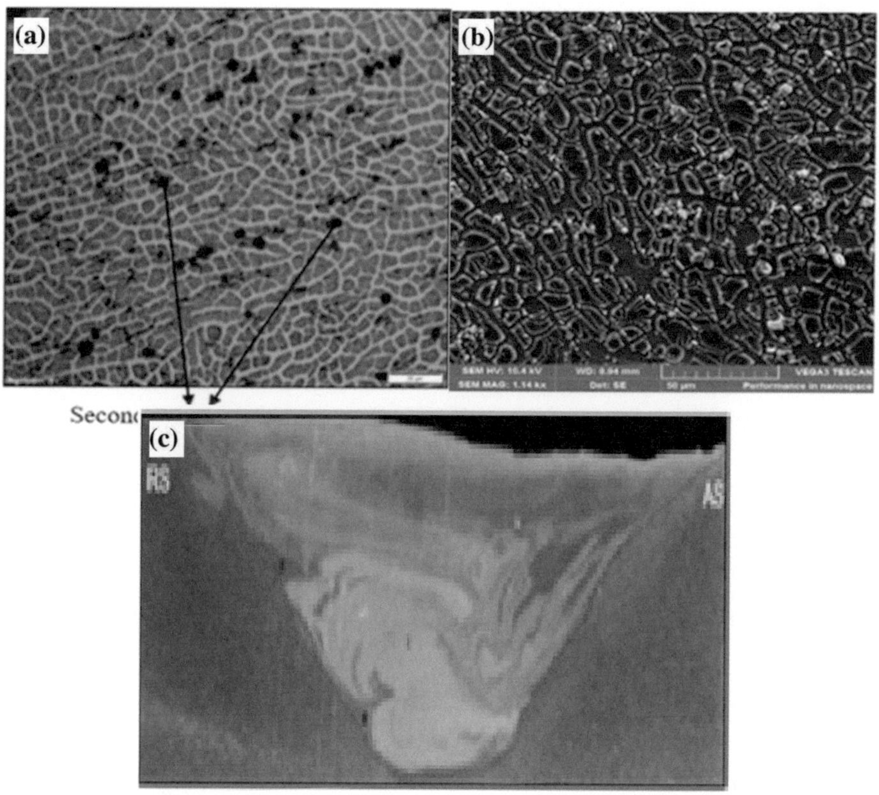

Fig. 6.3 Microstructure of the base material—AA1050. **a** OM **b** SEM **c** OM of the macrographs of FSPed samples

Table 6.2 Average grain size in various region of FPZ

Sample	SZ (μm)	TMAZ (μm)	HAZ (μm)
Al-FSPed	8.47	7.34	6.77

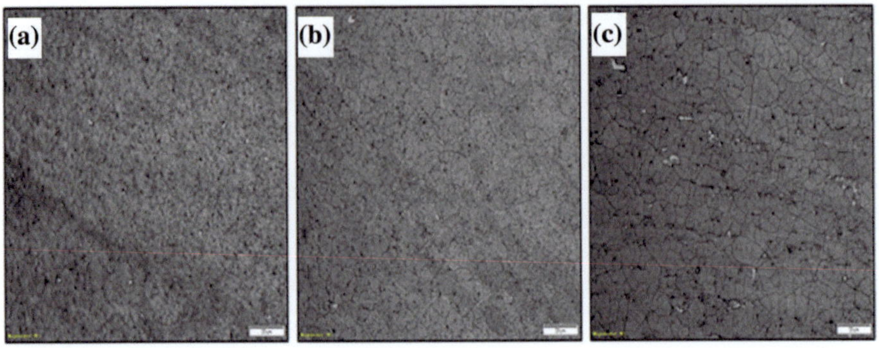

Fig. 6.4 OM micrographs of FSPed samples. **a** SZ, **b** TMAZ, **c** HAZ

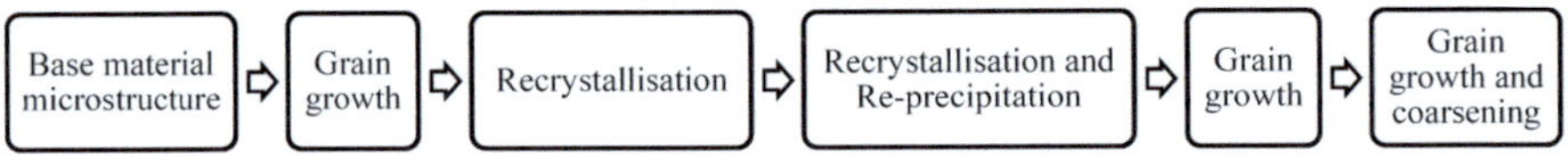

Fig. 6.5 Microstructural evolution procedure for FSP of Pure Aluminium

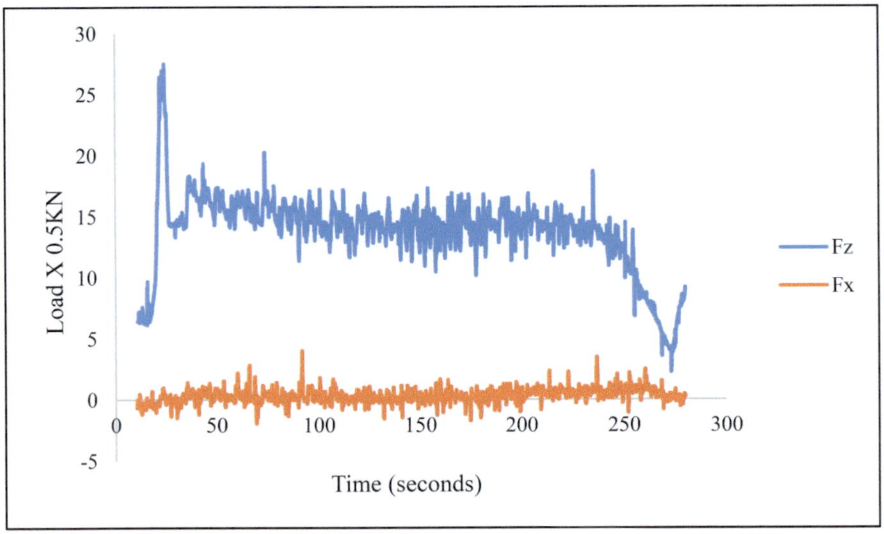

Fig. 6.6 Plot of force versus time (Plunging (axial) Force, Fz and Frictional force, Fx)

Table 6.3 Ultimate tensile strength of the base material and processed samples

Sample	T1 (MPa)	T2 (MPa)	T3 (MPa)	Mean UTS (MPa)
BM	84.49	84.79	85.83	85.04
FSPed	87.37	88.95	88.96	88.04

Table 6.4 Yield strength of the base material and processed samples

Sample	T1 (MPa)	T2 (MPa)	T3 (MPa)	Mean UTS (MPa)
BM	73.9	73.5	72.0	73.1
FSPed	77.4	76.8	77.7	77.3

Table 6.5 Elongation percentage of the base material and processed samples

Sample	Approximate % Elongation
BM	67.08
FSPed	69.21

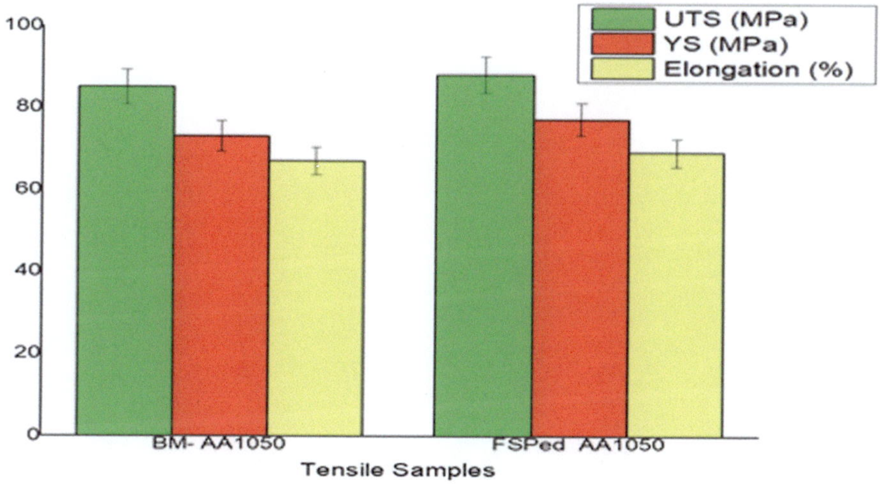

Fig. 6.7 Tensile values indicating UTS, YS and Percentage Elongation

of the tensile values is presented in Fig. 6.7 while Fig. 6.8 depicted SEM images of the fracture surfaces at lower and higher magnifications for both processed and unprocessed AA1050.

6.3.1 *Microstructural Examination*

An optical microscopy examination was carried out to study the influence of the FSP on the microstructure of FPZ. Figure 6.3a and b illustrates typical micrographs of the base material before FSP. The base material has a honeycomb-like microstructure with spherical shaped grains and relatively equiaxed grain size. Clearly, visible in homogenously distributed second-phase precipitates were seen within the entire

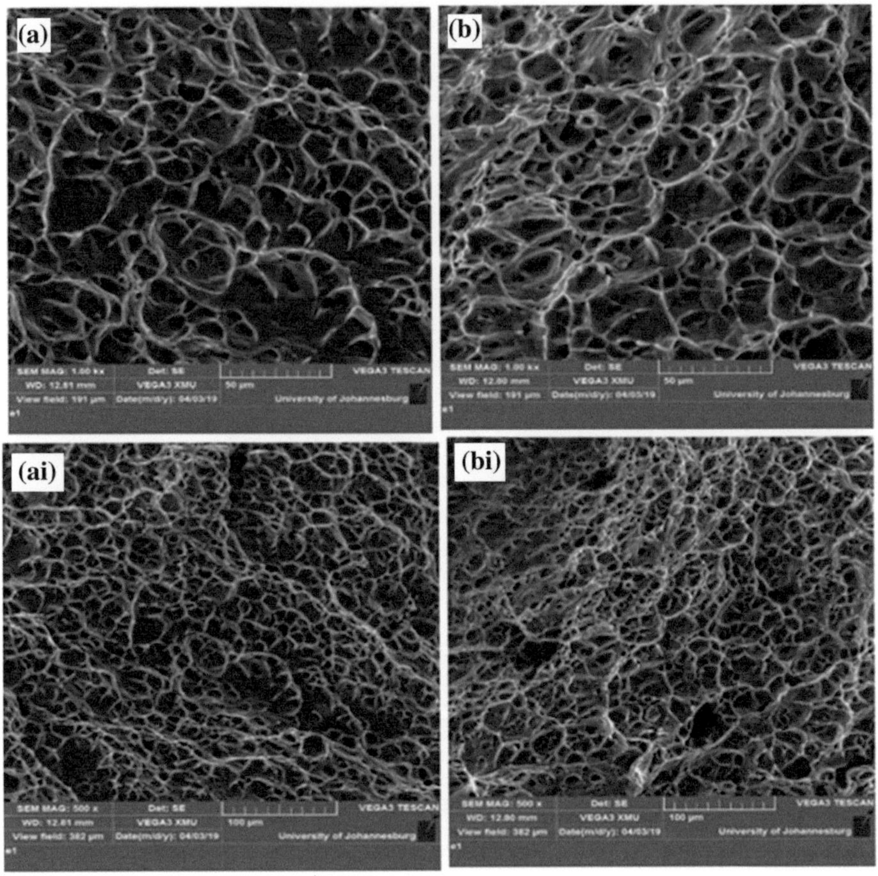

Fig. 6.8 SEM images of the fractured surfaces at 1000 MAG for **a** Base Metal—AA1050 **b** FSPed AA1050; at 500 MAG for **ai** Base Metal—AA1050 **bi** FSPed AA1050

base material microstructure matrix. Numerous studies have reported that the main second-phase strengthening precipitate found in pure aluminium.

Macrographs of FSPed samples are shown in Figs. 6.3c revealed that no internal defect is present in the FSPed samples. The absence of internal defects can be correlated to the process parameters and relatively stable axial force obtained in the force feedback. Mahto et al. [44] report that the forces and torque in FSW can be considered online indicators of weld quality. Their values during processing should be in a safe and stable range to obtain defect-free welds. It has been established that high axial forces may result in shear lips or flashes causing metal thinning at the processed area, while low axial forces lead to poor material consolidation due to insufficient forging pressure and friction heating. This is confirmed in a report by Abegunde et al. [45] that a high degree of variation in the Fz observed in a weld produced on 1050 alloy sheets of 300 × 200 × 3 mm at (1600–2000 rpm; 100–300 mm/min), produced a better-quality weld. In the force feedback from the FSP sample pass, the processing forces were observed to be very high during penetration of the tool as expected due to shearing but tended to decrease as there was sufficient frictional heat generated. The forces then remain almost constant throughout the process. These are similar to forces obtained in studies by Chen and Kovacevic [46] and Elangovan et al. [47] which are reported to have produced defect-free welds. The micrographs show two clearly distinguished zones; one with onion-ring-like structure and the remaining unmixed region which has been labelled region I and region II respectively, similar to the result by Patel et al. [48]. The shoulder-driven flow can be observed on the top portion of the macrographs while the probe-driven flow is seen towards the bottom part. Several researchers [49–51] have described the material flow as a combined process of pin-driven extrusion and shoulder-driven material stirring and mixing. The differences in the volume of microstructural homogeneity is a result of the differences in metal flow in the FSP samples processed. Region I depicts the area of the FPZ where the stirring action is severe while region II shows the unmixed area of the FPZ. A sharp boundary was only visible on the AS until the FPZ is completely homogenous. It was clear that region I increase in volume with an increase in the number of FSP passes towards the RS of the samples, significantly widening the basin-like shaped FPZ towards the upper surface due to the forging effect of the tool shoulder on material flow during FSP. This shows that effective stirring increases the microstructural homogeneity of materials, confirming the suggestion by Kumar and Kailas [52] and other studies [6, 53] that a single-pass FSP is not sufficient to attain microstructural homogeneity in the FSP of most materials. In the micrographs presented in Fig. 6.4, the three distinct FSP zones (SZ, TMAZ, HAZ) were studied. Microstructures in the different zones of the samples were carefully observed with OM to reveal the microstructural evolution after successive FSP. It can be observed that the microstructure of the FSPed regions is different from that of the base material. The FSPed zones exhibited a much more distinct spherical grain morphology compared to the base material. The SZ of the Al-sample consists of nearly equiaxed grains as a result of dynamic recrystallization. Grain growth and complete dissolution of the second-phase precipitate into the matrix is also clearly visible in the micrograph. The average grain size was found to be 6.69 μm using an intercept procedure [54], while the average size of the precipitates in the base material was found to be 6.57 μm.

The complete dissolution of the second-phase precipitates could be attributed to the inability of the identified precipitates to withstand high temperatures. Ahmed et al. [55] reported that the precipitates are not temperature resistant and rapidly dissolve when exposed to high temperatures resulting from FSP. Sanusi and Akinlabi [35] in their study of the influence of titanium carbide on the microstructural development during FSP on AA1050 also reported that the frictional heating resulting from FSP had caused the dissolution of the precipitates. Kovacs and Hareancz [56] reported that during FSP, the SZ experiences intense plastic deformation and thermal exposure with peak temperatures almost up to the melting point of the alloy. The size of the grains in the SZ, TMAZ and HAZ after the double passes are presented in Table 6.2.

A significant coarsening and growth of the grains are seen in the sample FPZ zone; the SZ seems to be coarser than the TMAZ and HAZ. This coarsening could be attributed to the additional/accumulated thermal cycles which the plate has experienced. Wang et al. [57] state a similar reason for grain coarsening. Pabandi et al. [58] also suggested that there may be some grain coarsening due to the additional thermal cycles on the plate. It is apparent that the various mechanisms acting at different stages of the microstructure evolution after every successive FSP pass are related to the strain, strain rate, and thermal cycle which the material undergoes at each stage. A summary of the mechanisms observed across the FSP pass micrographs is given in Fig. 6.5. Unfortunately, the thermal cycle experienced by the samples during each FSP pass was not quantitatively measured during this study; it would have been very interesting to study the relationship between the thermal cycles and the resulting microstructural evolutions. It is noteworthy that the resulting microstructural evolutions exercise significant influence on the mechanical properties of the FSPed samples. The tensile and hardness properties of the base material and the FSPed samples were studied and reported. The force feedback variations on the friction stir processed Aluminium alloy AA1050 showing the plunging force (axial) representing Fz and the frictional force representing Fx is depicted in Fig. 6.6.

Tables 6.3, 6.4 and 6.5 represent the tensile properties results for pure aluminium AA1050 for both base metal (BM) and the FSPed AA1050. It was established that FSPed AA1050 has improved tensile properties. It was established that average UTS values for FSPed AA1050 were higher than that of base metal (BM) by 3.42% leading to FSPed AA1050 having 88.04 MPa while BM has 85.04 MPa, this was in agreement with the work of Krishnaiah et al. [59]. A similar trend was also noticed in the yield strength (YS) whereby FSPed AA1050 has 77.3 MPa while BM has 73.1 MPa this led to a 5.4% increase by the FSPed AA1050. The percentage elongation was slightly higher in FSPed AA1050 than in base metal having 69.21–67.08%, respectively. The summary for all the properties, UTS, YS and percentage elongation is presented in Fig. 6.7.

6.3.2 *Fracture Mechanism*

The fracture behaviours of aluminium metal matrix composites (Al-MMC) has been traced to have been influenced by a couple of factors which includes type of materials (workpiece) used, volume and type of reinforcement applied, distributions of the reinforcement particles, the size of the particles whether is in nano, micro or macro, as well as the nature of matrix and interface properties—these may comprise of surface roughness, interfacial bonding strength, precipitations effect, as well as porosity content [60], etc. It is worth mentioning that particulate reinforcement failure in AMC has been traced to three different sources and they are reinforcement fracture, failure in the matrix as well as interfacial decohesion in the reinforcement matrix [61]. The tensile test was carried out for base metal—AA1050 and FSPed AA1050. The tensile samples failed at the centre for AA1050 and this can be attributed to the lower ultimate tensile strength possess as compared to the FSPed AA1050 that failed at base metal part (neck) of the processed samples. The no failure of the material at the nugget zone established that proper material mixing and bonding occurred considerably within the range of rotational and travel speeds deployed for this study. The fractography is taken and shown in Fig. 6.8a, b, ai and bi indicate that the fractured surfaces of processed AA1050 and unprocessed AA1050 exhibit similarities to a very large extent because the material fractured almost at the same point. The fractured samples indicate that there was appreciable plastic deformation and necking before failure took place. The fractured analysis equally shows fibrous features with appreciable lip region at the outer boundary of the final fracture. These noticeable fibrous features are an indication that the crack propagation occurred slowly. The cracking pattern exhibited what is termed equiaxed dimples with prominent cup and cone features, typical behaviour of ductile materials.

6.4 Summary

In this chapter, a case study of friction stir processing of pure aluminium has been presented. From the analysis of the experimental results, it can be concluded that tensile properties of friction stir processed (FSPed) AA1050 were greatly improved and enhanced; and this was also manifested in the fractography presented wherein fibrous-like of equiaxed dimples with prominent cup and cone features were noticed, indicating that the FSPed AA1050 was ductile in nature.

References

1. S. Rathee, S. Maheshwari, A.N. Siddiquee, M. Srivastava, Effect of tool plunge depth on reinforcement particles distribution in surface composite fabrication via friction stir processing. Def. Technol. **13**, 86–91 (2017). https://doi.org/10.1016/j.dt.2016.11.003
2. A.A. Cerit, M.B. Karamiş, N. Fehmi, Y. Kemal, Effect of reinforcement particle size and volume fraction on wear behaviour of metal matrix composites. Tribol. Ind. **30**, 31–36 (2008)
3. N. Sun, D. Apelian, Friction stir processing of aluminum cast alloys for high performance applications. Alum. Form. 44–50 (2011)
4. S.A. Nithin Reddy, R. Sathiskumar, K.G. Kumar, S. Jerome, A.V. Jebaraj, Friction based joining process for high strength aerospace aluminium alloy. Mater. Res. Express. **6**(8), 1–12 (2019). https://doi.org/10.1088/2053-1591/ab220c
5. P. Vijaya Kumar, G. Madhusudhan Reddy, R.K. Srinivasa, Microstructure and pitting corrosion of armor grade AA7075 aluminum alloy friction stir weld nugget zone—effect of post weld heat treatment and addition of boron carbide. Def. Technol. **11**, 166–173 (2015). https://doi.org/10.1016/j.dt.2015.01.002
6. T. Navaneethakrishnan, P. Ganesh, Effect of welding parameters on friction stir welded dissimilar aluminum alloys 7075 and 6082 with various tool pin profiles. J. Chem. Pharm. Sci. **2015**, 463–468 (April 2015)
7. A. Adetunla, E. Akinlabi, Significantly improved mechanical properties of 1100 aluminium alloy via particle reinforcement, in *5th International Conference Innovation Science Technology*, Barcelona, Spain, pp. 79–87 (2018)
8. P. Rambabu, N.E. Prasad, V.V. Kutumbarao, Aerosp. Mater. Mater. Technol. (2017). https://doi.org/10.1007/978-981-10-2143-5

9. J. Fayomi, A.P.I. Popoola, O.P. Oladijo, O.M. Popoola, O.S.I. Fayomi, Experimental study of ZrB_2-Si_3N_4 on the microstructure, mechanical and electrical properties of high grade AA8011 metal matrix composites. J. Alloys Comp. **790**, 610–615 (2019). https://doi.org/10.1016/j.jallcom.2019.03.112
10. C. Garcia, Z. Azpilgain, A. Aginagalde, J.A. Esnaola, L. Galdos, I. Torca, Tensile behaviour of 6082 aluminium alloy sheet under different conditions of heat treatment, temperature and strain rate. Key Eng. Mater. **423**, 105–112 (2009). https://doi.org/10.4028/www.scientific.net/kem.423.105
11. M. Vasco, K. Tserpes, S. Pantelakis, Numerical simulation of tensile behavior of corroded aluminum alloy 2024 T3 considering the hydrogen embrittlement. Metals (Basel) **8**, 56 (2018). https://doi.org/10.3390/met8010056
12. I. Sudhakar, G. Madhusudhan Reddy, K. Srinivasa Rao, Ballistic behavior of boron carbide reinforced AA7075 aluminium alloy using friction stir processing—an experimental study and analytical approach. Def. Technol. **12**, 25–31 (2016). https://doi.org/10.1016/j.dt.2015.04.005
13. M. Zohoor, M.K. Besharati Givi, P. Salami, Effect of processing parameters on fabrication of Al–Mg/Cu composites via friction stir processing. Mater. Des. **39**, 358–365 (2012). https://doi.org/10.1016/j.matdes.2012.02.042
14. R.M. Miranda, T.G. Santos, J. Gandra, N. Lopes, R.J.C. Silva, Reinforcement strategies for producing functionally graded materials by friction stir processing in aluminium alloys. J. Mater. Process. Technol. **213**, 1609–1615 (2013). https://doi.org/10.1016/j.jmatprotec.2013.03.022
15. O.M. Ikumapayi, E.T. Akinlabi, Efficacy of α-β grade titanium alloy powder (Ti–6Al–2Sn–2Zr–2Mo–2Cr–0.25Si) in surface modification and corrosion mitigation in 3.5% NaCl on friction stir processed armour grade 7075-T651 aluminum alloys—insight in defence applications. Mater. Res. Express. **2019** (2019). https://doi.org/10.1088/2053-1591/ab1566
16. N. Gangil, S. Maheshwari, A.N. Siddiquee, Surface nanocomposite fabrication on AA6063 aluminium alloy using friction stir processing: an investigation into the effect of the tool-shoulder diameter on the composite microstructure. Mater. Tehnolnol. **52**, 77–82 (2018). https://doi.org/10.17222/mit.2017.172
17. I. Dinaharan, E.T. Akinlabi, Low cost metal matrix composites based on aluminum, magnesium and copper reinforced with fly ash prepared using friction stir processing. Compos. Commun. **9**, 22–26 (2018). https://doi.org/10.1016/j.coco.2018.04.007
18. G. Hussain, R. Hashemi, Fabrication of metal-matrix AL7075T651/TiN nano composite employing friction stir process (2015). https://doi.org/10.1177/0954405415596695
19. K.A. Al-Ghamdi, G. Hussain, R. Hashemi, Fabrication of metal-matrix AL7075T651/TiN nano composite employing friction stir process. Proc. Inst. Mech. Eng. Part B J. Eng. Manuf. **231**, 1319–1331 (2017). https://doi.org/10.1177/0954405415596695
20. M. Sudhakar, C.H. Srinivasa Rao, K. Meera Saheb, Production of surface composites by friction stir processing. Mater. Today Proc. **5**, 929–935 (2018). https://doi.org/10.1016/j.matpr.2017.11.167
21. A. Bawdekar, S. Singh, Surface hardening and microstructural evolution of high carbon steel by friction stir processing. Int. J. Mech. Prod. Eng. Res. Dev. **7**, 343–354 (2017)
22. V. Sharma, U. Prakash, B.V.M. Kumar, Surface composites by friction stir processing: a review. J. Mater. Process. Technol. **224**, 117–134 (2015). https://doi.org/10.1016/j.jmatprotec.2015.04.019
23. I. Dinaharan, K. Kalaiselvan, E.T. Akinlabi, J.P. Davim, Microstructure and wear characterization of rice husk ash reinforced copper matrix composites prepared using friction stir processing. J. Alloys Comp. **718**, 150–160 (2017). https://doi.org/10.1016/j.jallcom.2017.05.117
24. K. Kumar, P. Gulati, A. Gupta, D.K. Shukla, A review of friction stir processing of aluminium alloys using different types of reinforcements. Int. J. Mech. Eng. Technol. **8**, 1638–1651 (2017)
25. D.K. Koli, G. Agnihotri, R. Purohit, A review on properties, behaviour and processing methods for Al- Nano Al_2O_3 composites. Procedia Mater. Sci. **6**, 567–589 (2014). https://doi.org/10.1016/j.mspro.2014.07.072
26. H. Das, M. Mondal, S.-T. Hong, D.-M. Chun, H.N. Han, Joining and fabrication of metal matrix composites by friction stir welding/processing. Int. J. Precis. Eng. Manuf. Technol. **5**, 151–172 (2018). https://doi.org/10.1007/s40684-018-0016-7

27. R.S. Mishra, Z.Y. Ma, Friction stir welding and processing. Mater. Sci. Eng. R Rep. **50**, 1–78 (2005). https://doi.org/10.1016/j.mser.2005.07.001
28. O.M. Ikumapayi, E.T. Akinlabi, J.D. Majumdar, Review on thermal, thermo-mechanical and thermal stress distribution during friction stir welding. Int. J. Mech. Eng. Technol. **9**, 534–548 (2018)
29. O.M. Ikumapayi, E.T. Akinlabi, A comparative assessment of tensile strength and corrosion protection in friction stir processed AA7075-T651 matrix composites using fly ashes nanoparticles as reinforcement inhibitors in 3.5% NaCl. Int. J. Mech. Prod. Eng. Res. Dev. **9**, 839–854 (2019). https://doi.org/10.24247/ijmperdjun201992
30. A.S. Kumar, V. Praveen, A review on aluminium based metal matrix composites by friction stir processing. Int. J. Eng. Manuf. Sci. **7**, 203–224 (2017)
31. R. Casati, M. Vedani, Metal matrix composites reinforced by nano-particles—a review. Metals (Basel) **4**, 65–83 (2014). https://doi.org/10.3390/met4010065
32. R.S. Mishra, P.S. De, N. Kumar, Frict. Stir Weld. Process. (2014). https://doi.org/10.1007/978-3-319-07043-8
33. A.P. Gerlich, Criticala Assessment 25: friction stir processing, potential and problems critical assessment. Mater. Sci. Technol. **33**, 1139–1144 (2017). https://doi.org/10.1080/02670836.2017.1300420
34. I. Dinaharan, Influence of ceramic particulate type on microstructure and tensile strength of aluminum matrix composites produced using friction stir processing. J. Asian Ceram. Soc. **4**, 209–218 (2016). https://doi.org/10.1016/j.jascer.2016.04.002
35. K.O. Sanusi, E.T. Akinlabi, Friction-stir processing of a composite aluminium alloy (AA 1050) reinforced with titanium carbide powder. Mater. Tehnol. **51**, 427–435 (2017). https://doi.org/10.17222/mit.2016.021
36. N. Fatchurrohman, N. Farhana, C.D. Marini, Investigation on the effect of friction stir processing parameters on micro-structure and micro-hardness of rice husk ash reinforced Al6061 metal matrix composites. IOP Conf. Ser. Mater. Sci. Eng. **319**, 0–6 (2018). https://doi.org/10.1088/1757-899X/319/1/012032
37. M. St. Węglowski, Friction stir processing technology—new opportunities. Weld. Int. **28**, 583–592 (2014). https://doi.org/10.1080/09507116.2012.753216
38. S. Lin, Y.L. Deng, H.Q. Lin, L.Y. Ye, Z. Zhang, H. Ji et al., Microstructure, mechanical properties and stress corrosion behavior of friction stir welded joint of Al–Mg–Si alloy extrusion. Rare Met. 1–11 (2018). https://doi.org/10.1007/s12598-018-1126-7
39. Y.X. Gan, D. Solomon, M. Reinbolt, Friction stir processing of particle reinforced composite materials. Materials (Basel) **3**, 329–350 (2010). https://doi.org/10.3390/ma3010329
40. M.C. Pereira, J.W.J. Silva, H.A. Acciari, E.N. Codaro, L.R.O. Hein, Morphology characterization and kinetics evaluation of pitting corrosion of commercially pure aluminium by digital image analysis. Mater. Sci. Appl. **3**, 287–293 (2012). https://doi.org/10.4236/msa.2012.35042
41. O.M. Ikumapayi, I.P. Okokpujie, S.A. Afolalu, O.O. Ajayi, E.T. Akilabi, O.P. Bodunde, Effects of quenchants on impact strength of single-vee butt welded joint of mild steel. IOP Conf. Ser. Mater. Sci. Eng. **391**, 1–10 (2018). https://doi.org/10.1088/1757-899X/391/1/012007
42. O.M. Ikumapayi, S.J. Ojolo, S.A. Afolalu, Experimental and theoretical investigation of tensile stress distribution during aluminium wire drawing. Eur. Sci. J. **11**, 1857–7881 (2015). https://doi.org/10.1016/j.jmatprotec.2005.02.122
43. X.W. Yang, T. Fu, W.Y. Li, Friction stir spot welding: a review on joint macro-and microstructure, property, and process modelling. Adv. Mater. Sci. Eng. **2014**, 1–11 (2014)
44. R.P. Mahto, R. Kumar, S.K. Pal, S.K. Panda, A comprehensive study on force, temperature, mechanical properties and micro-structural characterizations in friction stir lap welding of dissimilar materials (AA6061-T6 & AISI304). J. Manuf. Process. **31**, 624–639 (2018). https://doi.org/10.1016/j.jmapro.2017.12.017
45. O.O. Abegunde, E.T. Akinlabi, D.M. Madyira, Force feedback analysis: friction stir welding of aluminium metal matrix composite. Lect. Notes Eng. Comput. Sci. **2224**, 1–6 (2016)
46. C. Chen, R. Kovacevic, Thermomechanical modelling and force analysis of friction stir welding by the finite element method. Proc. Inst. Mech. Eng. Part C J. Mech. Eng. Sci. **218**, 509–520 (2004). https://doi.org/10.1243/095440604323052292

47. K. Elangovan, V. Balasubramanian, M. Valliappan, Influences of tool pin profile and axial force on the formation of friction stir processing zone in AA6061 aluminium alloy. Int. J. Adv. Manuf. Technol. **38**, 285–295 (2008). https://doi.org/10.1007/s00170-007-1100-2
48. V.V. Patel, V. Badheka, A. Kumar, Effect of polygonal pin profiles on friction stir processed superplasticity of AA7075 alloy. J. Mater. Process. Technol. **240**, 68–76 (2017). https://doi.org/10.1016/j.jmatprotec.2016.09.009
49. T.K. Bhattacharya, H. Das, T.K. Pal, Influence of welding parameters on material flow, mechanical property and intermetallic characterization of friction stir welded AA6063 to HCP copper dissimilar butt joint without offset. Trans. Nonfer. Met. Soc. China (English Ed) **25**, 2833–2846 (2015). https://doi.org/10.1016/s1003-6326(15)63909-7
50. E. Hoyos, D. López, H. Alvarez, A phenomenologically based material flow model for friction stir welding. Mater. Des. **111**, 321–330 (2016). https://doi.org/10.1016/j.matdes.2016.09.009
51. H. Robe, Y. Zedan, J. Chen, H. Monajati, E. Feulvarch, P. Bocher, Microstructural and mechanical characterization of a dissimilar friction stir welded butt joint made of AA2024-T3 and AA2198-T3. Mater. Charact. **110**, 242–251 (2015). https://doi.org/10.1016/j.matchar.2015.10.029
52. K. Kumar, S.V. Kailas, The role of friction stir welding tool on material flow and weld formation. Mater. Sci. Eng., A **485**, 367–374 (2008). https://doi.org/10.1016/j.msea.2007.08.013
53. E. Moustafa, Effect of multi-pass friction stir processing on mechanical properties for AA2024/Al_2O_3 nanocomposites. Materials (Basel) **10**, 1053 (2017). https://doi.org/10.3390/ma10091053
54. S. Rajakumar, V. Balasubramanian, Predicting grain size and tensile strength of friction stir welded joints of AA7075-T_6 aluminium alloy. Mater. Manuf. Process. **27**, 78–83 (2012). https://doi.org/10.1080/10426914.2011.557123
55. A.A.G.A. Ahmed, et al., Process parameters optimization of friction stir processed Al 1050 aluminum alloy by response surface methodology (RSM). Mater. Res. Express. 1–32 (2018). https://doi.org/10.1088/2053-1591/aaed7c
56. Z. Kovács, F. Hareancz, Joining of non-weldable AA7075 and weldable AA6082 aluminium alloy sheets by friction stir welding. IOP Conf. Ser. Mater. Sci. Eng. **448**, 012001 (2018). https://doi.org/10.1088/1757-899X/448/1/012001
57. K. Wang, F.C. Liu, P. Xue, B.L. Xiao, Z.Y. Ma, Effects of heating rates on microstructure and superplastic behavior of friction stir processed 7075 aluminum alloy. J. Mater. Sci. **50**, 1006–1015 (2014). https://doi.org/10.1007/s10853-014-8660-y
58. H.K. Pabandi, H.R. Jashnani, M. Paidar, Effect of precipitation hardening heat treatment on mechanical and microstructure features of dissimilar friction stir welded AA2024-T6 and AA6061-T6 alloys. J. Manuf. Process. **31**, 214–220 (2018). https://doi.org/10.1016/j.jmapro.2017.11.019
59. A. Krishnaiah, U. Chakkingal, H.S. Kim, Mechanical properties of commercially pure aluminium subjected to repetitive bending and straightening process. Trans. Indian Inst. Met. **61**, 165–167 (2008). https://doi.org/10.1007/s12666-008-0032-3
60. H. Li, P. Chen, Z. Wang, F. Zhu, R. Song, Z. Zheng, A Tensile properties, microstructures and fracture behaviors of an Al–Zn–Mg–Cu alloy during ageing after solution treating and cold-rolling. Mater. Sci. Eng., A **742**, 798–812 (2019). https://doi.org/10.1016/j.msea.2018.03.098
61. Z. Hao, X. Fu, X. Men, B. Zhou, Study on tensile and fracture properties of 7050-T7451 aluminum alloy based on material forming texture characteristics. Mater. Res. Express. **6** (2019). https://doi.org/10.1088/2053-1591/aaf304

Chapter 7
Future Research Direction in Friction Welding, Friction Stir Welding and Friction Stir Processing

7.1 Introduction

There is a constant demand for lightweight materials with conflicting properties that sometimes cannot be achieved in a single material which makes it necessary to weld two or more dissimilar materials together. Aerospace and automobile industries are interested in high fuel-efficient multi-material combinations with lightweight to reduce their carbon footprint. Different material combinations that offer varieties of properties include aluminium alloys, magnesium alloys, copper and steels. These dissimilar materials need to be joined which pose lots of challenges as a result of their incompatibilities and the conventional fusion welding processes cannot be used. Friction-based joining techniques have proved to be useful in joining such dissimilar materials and are capable of producing defect-free welds because they are solid-state processes and the problem of porosity defect and material segregation that are common to fusion welding are prevented. Extruded parts can readily be welded using friction welding and friction stir welding to produce larger extrusions as well as in joining various forms of assembly. For these solid-state welding techniques to be cost-effective and reliable for joining lightweight alloys, such as aluminium, and high-density materials, such as steel, will demand significant development and consideration. A lot has been done in terms of research and more still need to be done to take full advantage of these technologies. The future research needs in these solid-state welding processes are analysed in this chapter. The frictions stir processing of material surfaces as well as bulk material has been widely studied and with impressing results. Grain refinement and ultra-fine-grained structures are desired in many applications for improved properties and increasing the formability of materials with superplasticity. More research is still needed to further understand the process physics which will enable adequate process control and hence to achieve desired control of material properties. Some of these research needs are also presented in the

E. T. Akinlabi and R. M. Mahamood, *Solid State Welding: Friction and Friction Stir Welding Processes*, Mechanical Engineering Series,
https://doi.org/10.1007/978-3-030-37015-2_7

next section. The Fourth Industrial Revolution is here and the way products are made are rapidly changing, to position the friction stir welding for the Fourth Industrial Revolution, the role of cloud computing in friction stir welding is also presented.

7.2 Future Research Need in Friction Welding, Friction Stir Welding and Friction Stir Processing

The awareness on our environment is on the increase and the effect of man activities on the environment is evident with the global warming and other environmental impacts that these activities have on the environment. The need to reduce emission on the environment has been taken with all seriousness and has been backed up with legislation in different countries. The United State government and Corporate Average Fuel Economy (CAFÉ) program have mandated higher fleet efficiency. The International Council on Clean Transportation is also demanding a target of 95 g CO_2/km for vehicle emissions, which can be met by aggressive weight reduction of all the moving parts [1]. All these legislative restrictions have been the key driver for research on technological developments that include the development of high performing materials, improved material formability, ultra-high-strength material higher strength to weight ratios, etc. These materials can be used simultaneously to optimize the properties. The weldability of dissimilar material such as ultra-high-strength crash resistance materials is still a challenge. In order to achieve the required weight saving in the transport industries, dissimilar materials need to be welded instead of the use of fasteners that will add extra weight to the component. Friction-based welding processes offer an advantage in terms of weight saving because there is no filler material that can further add to the net weight. These solid-state welding processes are also environmental friendly. Friction welding and friction stir welding (FSW) have been favoured in the manufacturing and automotive industries [2].

Friction welding has been successfully employed to weld similar and dissimilar metals with good strength of the weld as a result of research advances in the field [3–8]. More still needs to be done in terms of research. There is a need to adequately model the friction welding process using finite element method, finite difference method etc. Availability of adequate model for the friction welding process will go a long way in how the process can be adequately controlled. More research work is also needed for more characterization of the weld produced during its service life. More research work is also needed to understand how friction welded materials fail in service. Proper fracture analysis is required to further understand the process. This is also important to further advance the technology further. More research is also needed in understanding the residual stress generation during the friction welding process and the residual stress development should also be modelled adequately. The effect of the friction welding machine on the weld properties needs to be thoroughly investigated. The right way to apply pressure during the welding process and if the applied pressure should be ramped up to the maximum value quickly or slowly needs to also be investigated. The best method to apply motion (oscillatory, sinusoidal, etc.) during the welding process should be optimized through adequate research. The influence of the workpiece clamping devices on the flash formation and interface contaminant removal needs to be investigated. The understanding of the effect of all these parameters on the resulting weld properties will go a long way in optimizing the friction welding process. Further studies are also needed to further understand the post-welding heat treatment and its effect on the reduction of residual stresses.

In friction stir welding of dissimilar materials with wide apart melting points could result in some melting of the material with a lower melting point even though FSW is a solid-state process [9–20]. This can result in the formation of intermetallic compound at the solid–liquid interface. One of the attractive feature of solid-state welding process is the temperature within the weld region can be controlled to prevent the formation of undesirable brittle intermetallic compounds within the weld that can adversely affect the strength of the weld. A number of researches have been carried out and reported in the academic literature, however, more research work is still needed to study the physical and mechanical properties of the weld zone. There is a need to further understand the influence of process parameters on the weld properties and performance. More geometries need to be developed and the influence of tool geometry on the weld properties and performances need to be thoroughly investigated. Recent research advancement has led to the development of tools that are now used to weld material with high melting points such as steel and titanium and its alloys. There is still a knowledge gap that needs to be bridged in order to aid further understanding of the process physics and also the wide use of the technology for the manufacturing of critical parts. Microstructure Modelling is another research area that needs to be further explored. Proper mapping of the microstructure and the processing parameters is required. With an adequate model, the properties of the friction welded part can be adequately predicted and it can aid in the proper design of engineering materials. This better knowledge of the influence of the processing parameters on the microstructural evolution such as average grain size, allotropic phase and grain spatial distribution in the weld will go a long way in property control during the friction stir welding process. This could also help to have a better understanding of the reasons for the different texture developments in the weld region. Further research is also needed to investigate different misalignment that can result from the FSW process and their effect on the fatigue behaviour. Scientific knowledge-based numerical studies are of needed to improve the current understanding of the FSW process. The friction stir welding of metal matrix composite materials need further investigation because there is no general trend between welding parameters and mechanical properties for different types of composites. The welding process window of each of the composites needs to be established. More in-depth research work is needed to have more fatigue data for friction stir welding process. New tool material is also needed to be developed so that the tool wear that is currently experienced can be reduced.

Friction stir processing has revolutionized the way surface properties are modified through microstructural modification and grain refinement. Friction stir processing is also used in producing composite on surface of materials [21–29]. More research work is needed to study the influence of different tool design and pin configuration on the evolving properties. Development of different tool design is required that is completely different from the one used in the friction stir welding process. Larger tools need to be designed that will be capable of wide coverage at a single pass. This will help to improve the performance of the process as well as improve the speed and productivity. This microstructural modification process can be further explored through research to develop novel materials through the study of a combination of two or more reinforcement materials and study their effect on the evolving properties.

7.3 Fourth Industrial Revolution and Cloud Computing in Friction Stir Welding

Fourth Industrial Revolution (4IR) also known as Industry 4.0 is an unprecedented and simultaneous advances in technologies such as the artificial intelligence (AI), the internet of things, robotics, 3D printing, nanotechnology, autonomous vehicles, biotechnology, materials science, energy storage, and cloud computing are redefining the industries, removing the traditional boundaries and also creating new opportunities. Industry 4.0 is fundamentally changing the way work is done and human is relating to machine vis-à-vis. Industrial revolutions have always been the bedrock of human civilization by transforming human ingenuity to innovations and human development. Looking back at the previous industrial revolutions, there is a common trend of how the industrial revolution evolved which is based on technological innovations that fundamentally changed society [30–32].

The First Industrial Revolution started in Britain in the 1760s. It occurred as a result of the wave of innovations. The development of the steam engine and the cotton mill caused a lot of change such as mechanization that replaced agriculture with industry creating a new structure of economic prospect for the society. The extraction of coal on a large scale and the invention of the steam engine created a new type of energy that helped to advance human development at that time. During the First Industrial Revolution, other inventions also evolved such as new technological know-how of metal shaping that aided in machine development for the creation of factories. The First Industrial Revolution saw the development of agricultural mechanization, mass education, industrialization, factory development and urbanization. The society at that time saw an unprecedented change in the way things are done and impacted human life as a whole.

The Second Industrial Revolution started in the 1870s about a century after the First Industrial Revolution at the end of the nineteenth century. The technological innovations that drove the Second Industrial Revolution include the emergence of new energy, electricity, oil and gas and the development of internal combustion engine. The new capability also led to the development of other technologies such as the development and growing of the steel industry as a result of the rising demands for steel, the development of synthetic fabric, dyes, fertilizer, the invention of the telegraph and the telephone, emergence of the automobile and the aeroplane that started at the beginning of the twentieth century. The Second Industrial Revolution, electric energy innovation made mass production possible and promoted the development of entirely new social, economic and overall new ways of life for the people at that time.

The Third Industrial Revolution began in the late 1960s at the mid of the twentieth century almost a century after the advent of the Second Industrial Revolution. The way of Third Industrial Revolution was brought about by the emergence of a new type of energy with potentials that surpassed its predecessors that is nuclear energy development. Also, the invention of the semiconductor, transistor and microprocessor led to the further development of telecommunications and the invention of personal computers. This new technology led to the rise of industrial automation due to the ability to produce miniaturized material. The invention of the Internet marked the peak of the Third Industrial Revolution and the Third Industrial Revolution is also called the Digital Revolution.

There has been rapid technological development and innovations since the evolution of the third industrial revolution. The gap between the digital, physical and biological worlds is constantly shrinking, and new technologies are evolving and springing up at an unprecedented rate. The first three industrial revolutions have defined the contours of the Fourth Industrial Revolution that is already happening. The Fourth Industrial Revolution also known as Industry 4.0 will not only change our industries as we know it but also going to change everything about our world as we know it from creating a new mental model to help businesses, the government and the at large. The Fourth Industrial Revolution is described as how technologies such as artificial intelligence, the internet of things, and autonomous vehicles, voice-activated assistants, facial ID recognition and digital healthcare sensors are impacting the humans lives at the moment. These technological advancements are drastically changing how individuals, businesses and governments operate that is generating a new wave of societal transformation that is similar to what happened during the previous industrial revolutions. The Fourth Industrial Revolution is characterized by merging different technologies that keep reducing the gap between the physical, the digital and the biological spheres that will completely change industries around the world. This industrial revolution will impact the entire production world, the management and governance systems. The foundation of the Fourth Industrial Revolution is based on the emergence of the Internet at the tail end of the Third Industrial Revolution. This time around, an industrial revolution is happening not because of the emergence of new energy but it is based on a new technological phenomenon which is called digitalization. The digitalization enables the building of a new virtual world from which the physical world can be controlled. The Fourth Industrial Revolution has begun, it is characterized by what is called "Cyber-Physical Systems" (CPS). Systems are integrated for higher productivity, sustainability and customer-satisfaction all these are leveraging on intelligent network systems and processes. The 'internet of things' is one of the key driving forces in this industrial revolution. The floor shops in industries are now connecting to the Internet. The development of microprocessors have paved ways for this to happen with the high acceleration of 'Internet of industry' made possible with the unprecedented development of digital devices. It is now easier to connect devices, appliances, machines, homes and even a complete factory and factories to the Internet with the advent of 'Industrial Internet Protocol (Industrial IP)' that was established in 2013. The objective of industrial IP is to enable all industrial network infrastructures and applications to communicate with one another and benefit from the end-to-end Internet connectivity coupled with the Common Industrial Protocol (CIP) that enables unified communications. With this information technological developments, machines and equipment, real-time control can be actualized with sensors and actuators from anywhere. This industrial IP enables the process and analyses of significantly larger information or data flow between the production and auxiliary processes, making it possible for companies to operate in a more flexible and innovative way. This is key for companies to remain competitive in this new Industry 4.0, the era of customization of goods and services. New businesses and services are now being developed to link the physical and virtual worlds together. The Industry 4.0 is aimed to connect all production facilities to enable them to interact in real time. Creating smartness of everything such as smart homes, smart factories, smart cities, etc. The industry Factories 4.0 makes the communication among the different components in a production line possible with the help of Cloud, Big Data Analytics and the Industrial Internet of Things. This has helped factories to improve productivities and reduce downtime through predictive maintenance, improved decision-making in real time, improved coordination among jobs, anticipating inventory based on production, etc. The Industry 4.0 is connecting information technology (IT) and operational technology (OT) together to achieve this phenomenal development. The OT is the hardware and software that detects or causes a change through the direct monitoring and/or control of physical devices—cyber-physical Internet-based systems. The machine-to-machine communication is now possible not only between machines in one factory but also between all devices and systems they are connected to all over the world. Sensors have a greater role to play in this industrial revolution as well as artificial intelligence. Some of the benefits of this Industry 4.0 include: Machine-to-machine communication that helped to reduce human work thereby increasing efficiency; maintenance are changing from corrective or scheduled maintenance to predictive maintenance of machines and appliances based on the real time data that gives the current status of the machine and determine when maintenance needs to be carried out and possibly remotely and consumers can also communicate with the product they are using such as a car telling the driver to check engine, etc. Now, all these devices and human can communicate with one another with the help of industrial internet and innovation. For example, domestic appliances such as washing machines, toothbrushes and the car will be able to tell the user things like I need more detergent, I have not been used this morning indicating that someone has forgotten to brush his or her teeth and car telling the driver that if certain percentage of more pressure is put in the tyres, the car can save fuel up to certain percentage. Cloud computing has a big role to play in Industry 4.0.

Cloud computing is a paradigm that allows on-demand network access to shared computing resources. It is a model that allows the storing and processing of data online using the Internet. The advancement in computational power of modern-day computer, communication, and large storage capability and the benefits of the cloud and related services, has resulted a new generation of service-oriented architecture (SOA)-based industrial systems with seamless communication between devices and the cloud [33] that has significantly impacted the manufacturing industries as well as other industries rendering services to its client [34, 35]. In the modern-day world, the consumer demand keeps changing and moving away from the massively produced product to more customized ones [36]. To remain competitive in the new manufacturing world, production processes are now performed in highly distributed production systems that need to be efficiently integrated with a sophisticated shop-floor infrastructure that is agile and flexible [37]. Figure 7.1 represents a typical architecture of industrial infrastructure in cyber-physical systems (CPS), that is based on service-oriented architecture and cloud computing [38, 39]. This will result in heterogeneous, dynamic and adequately performing ecosystem of services such as Supervisory Control And Data Acquisition (SCADA), Enterprise Resource Planning (ERP) Manufacturing Execution Systems (MES) [40].

For a manufacturing process to be competitive in the modern day, it needs to be Industry 4.0 compatible. Friction stir welding needs to be adequately positioned for this new industrial revolution that is already evolving. FSW need to achieve a higher level of the autonomous and centralized control system that is implemented using the cloud and edge/fog computing [41, 42]. The basic components of smart factory are: cloud computing platform: A cloud [43] is located on a remote server where processor, database, memory and all necessary computing resources that can be accessed by several units that are connected to the cloud through the Internet at any time are kept [44]. Fog/Edge computing platform: The fog layer is a decentralized edge processing device helps to improve the efficiency of the internet of things based network [45]. Digital twin: It uses data processing, machine learning and software

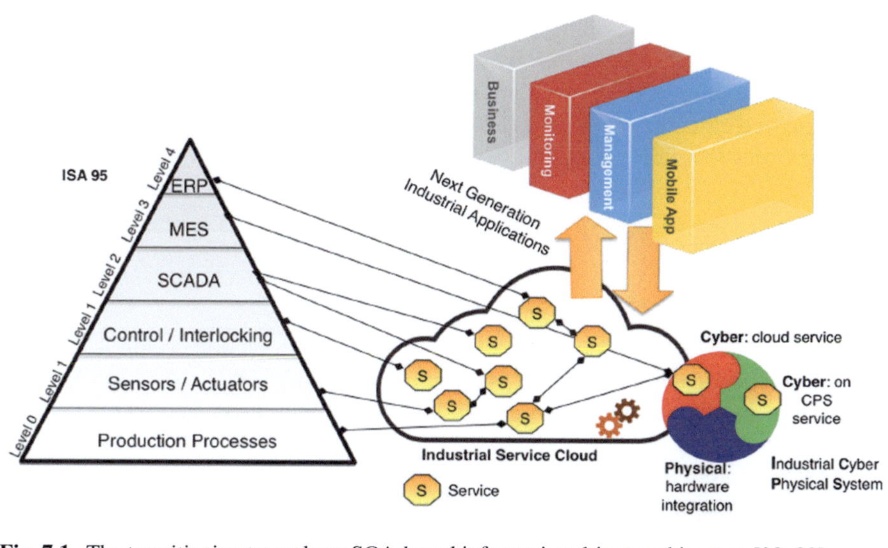

Fig. 7.1 The transitioning towards an SOA-based information-driven architecture [38, 39]

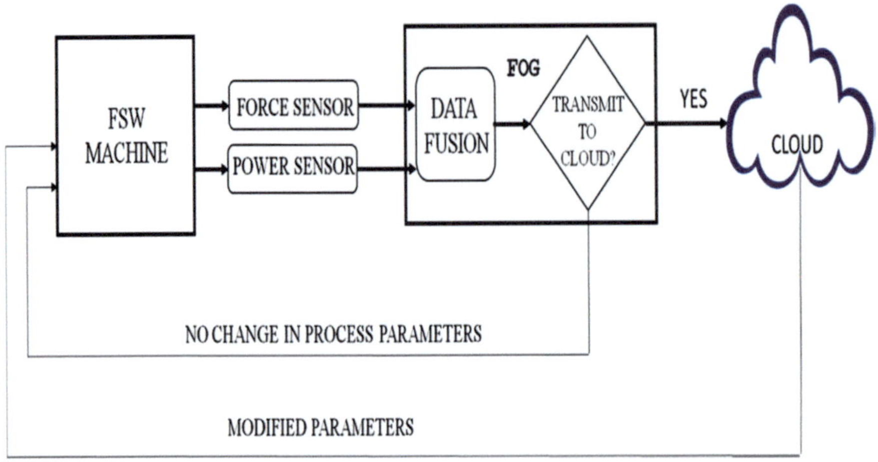

Fig. 7.2 Fog–cloud integration for an adaptive control of FSW [42]

analytics to create a sensor mapped digital simulation of a physical system for data collection [46]. Cloud-fog-based remote manufacturing and process parameter control of FSW, and, continuous monitor machine health and take corrective measures to restore machine health [42]. Each of the FSW machine parts will also have sensors connected to them for integration to the cloud. These sensors include force, torque, power, vibration, etc., and data acquisition unit for data collection. The collected data will be transmitted to the fog computing unit that is associated with each of the FSW machine parts, as shown in Fig. 7.2 [42]. The classification of data in friction stir welding for online and offline control are shown in Fig. 7.3 [42]. Mishra et al. [42] proposed a friction stir welding smart factory as shown in Fig. 7.4 with a monitoring and control architecture as shown in Fig. 7.5. Autonomous fully connected friction stir welding smart factory for Industrial 4.0 will reduce the manufacturing cost, improved efficiency, reduce material wastage and give a competitive advantage to the friction stir welding process.

7.4 Summary

In this chapter, the future research needs to be done in friction welding, friction stir welding and friction processes were presented. The need to position friction stir welding for the Fourth Industrial Revolution was presented. The previous three industrial revolutions and the Fourth Industrial Revolution were introduced. The role of the current innovations in the industry 4.0 was highlighted. Cloud computing is an important component of industry 4.0 and its importance in driving the friction stir welding as a smart factory was also highlighted.

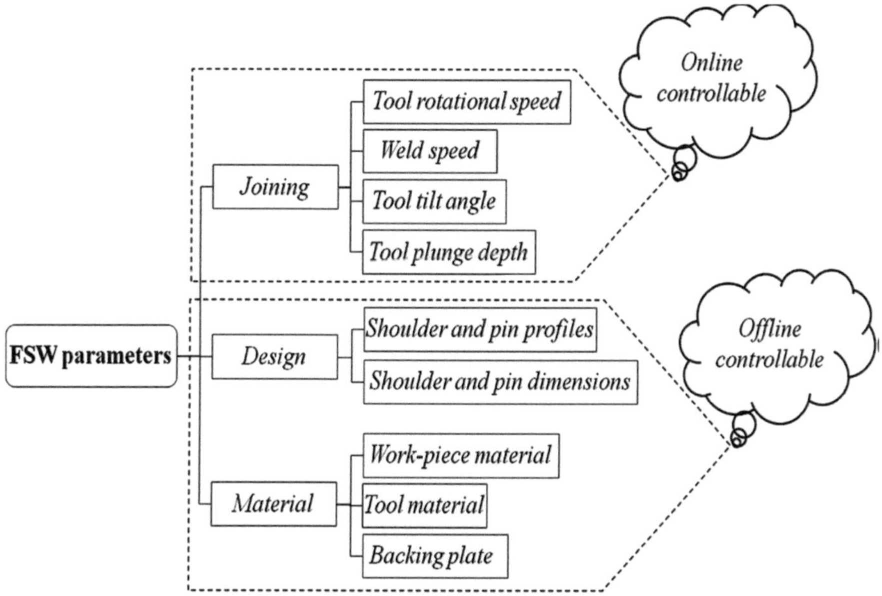

Fig. 7.3 Classification of parameters [42]

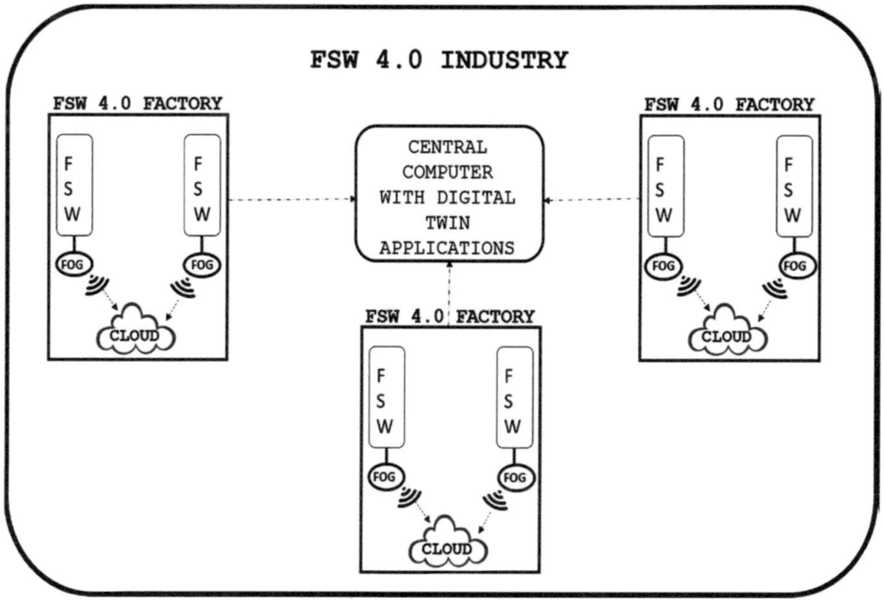

Fig. 7.4 FSW smart industry [42]

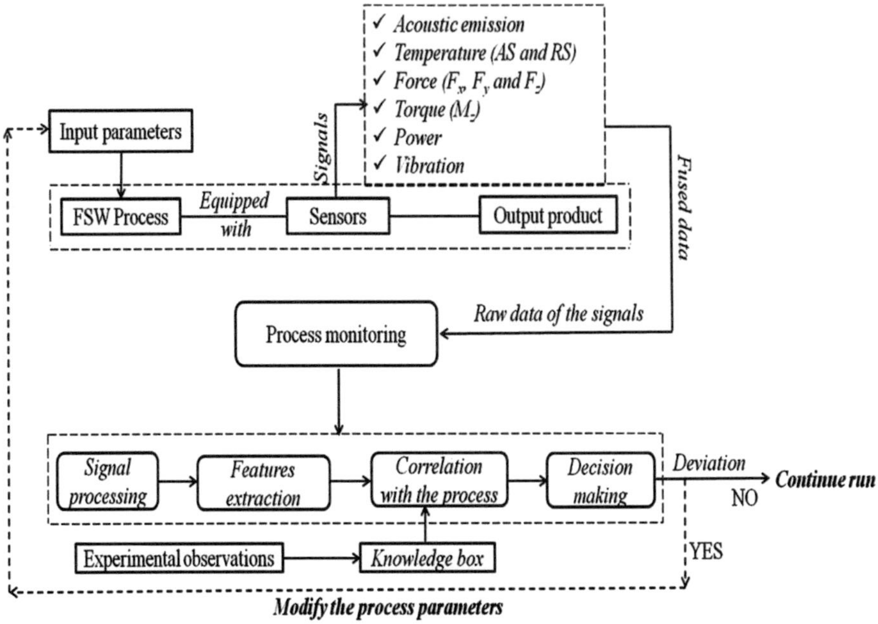

Fig. 7.5 FSW monitoring and control architecture [42]

References

1. R.S. Mishra, Z.Y. Ma, Mater. Sci. Eng., R **50**, 1–78 (2005)
2. P.L. Threadgill, A.J. Leonard, H.R. Shercliff, P.J. Withers, Int. Mater. Rev. **54**, 49–93 (2009)
3. F. Rotundo, A. Marconi, A. Morri, A. Ceschini, Dissimilar linear friction welding between a SiC particle reinforced aluminum composite and a monolithic aluminum alloy: microstructural, tensile and fatigue properties. Mater. Sci. Eng., A **559**, 852–860 (2013). https://doi.org/10.1016/j.msea.2012.09.033
4. M. Grujicic, G. Arakere, B. Pandurangan, C.F. Yen, B.A. Cheeseman, Process modeling of Ti-6Al-4V linear friction welding (LFW). J. Mater. Eng. Perform. **21**, 2011–2023 (2011). https://doi.org/10.1007/s11665-011-0097-8?ASM
5. M.Y. Xie, T.S. Jun, A.M. Korsunsky, M. Drakopoulos, Analysis of preferred orientations in linear friction welded (LFW) aluminium alloy specimens using "one-shot" multi-element energy dispersive synchrotron X-ray diffraction. Powder Diffr. **28**, 327–332 (2013). https://doi.org/10.1017/S0885715613000948
6. A. Chamanfar, M. Jahazi, J. Gholipour, P. Wanjara, S. Yue, Mechanical property and microstructure of linear friction welded WASPALOY. Metall. Mater. Trans. A **42**, 729–744 (2010). https://doi.org/10.1007/s11661-010-0457-2
7. P. Wanjara, E. Dalgaard, G. Trigo, C. Mandache, G. Comeau, J.J. Jonas, Linear friction welding of Al–Cu: Part 1—Process evaluation. Can. Metall. Q. **50**, 350–359 (2011). https://doi.org/10.1179/000844311X13112418194644
8. J.P. Bergmann, F. Petzoldt, R. Schürer, S. Schneider, Solid-state welding of aluminum to copper—case studies. Weld. World **57**, 541–550 (2013). https://doi.org/10.1007/s40194-013-0049-z
9. J.D. Robson, S. Cui, Z.W. Chen, Mater. Sci. Eng., A **527**, 7299–7304 (2010)

10. T.H. North, G.J. Bendzsak, A. Gerlich, P. Su, G. Cingara, Mater. Sci. Forum **539–543**, 3826–3831 (2007)
11. S. Papaefthymiou, C. Goulas, E. Gavalas, Micro-friction stir welding of titan zinc sheets. J. Mater. Process. Technol. **216**, 133–139 (2015). https://doi.org/10.1016/j.jmatprotec.2014.08.029
12. K. Panneerselvam, K. Lenin, Joining of Nylon 6 plate by friction stir welding process using threaded pin profile. Mater. Des. 302–307 (2014). http://dx.doi.org/10.1016/j.matdes.2013.07.017
13. H. Ahmadi, N.B. Mostafa Arab, F. Ashenai Ghasemi, R. Eslami Farsani, Influence of pin profile on quality of friction stir lap welds in carbon fiber reinforced polypropylene composite. Int. J. Mech. Appl. **2**(3), 24–28 (2012). http://dx.doi.org/10.5923/j.mechanics.20120203.01
14. I. Galvão, R.M. Leal, A. Loureiro, D.M. Rodrigues, Material flow in heterogeneous friction stir welding of aluminium and copper thin sheets. Sci. Technol. Weld. Join. 654–660 (2010). http://dx.doi.org/10.1179/136217110X12785889550109
15. A. Scialpi, L.A.C. De Filippis, P. Cuomob, P. Di Summa, Micro friction stir welding of 2024–6082 aluminium alloys. Weld. Int. 16–22 (2008). http://dx.doi.org/10.1080/09507110801936069
16. R.M. Leal, C. Leitão, A. Loureiro, D.M. Rodriguesa, P. Vilaça, Material flow in heterogeneous friction stir welding of thin aluminium sheets: effect of shoulder geometry. Mater. Sci. Eng. 384–391 (2008). http://dx.doi.org/10.1016/j.msea.2008.08.018
17. D.M. Rodrigues, A. Loureiro, C. Leitão, R.M. Leal, B.M. Chaparro, P. Vilaça, Influence of friction stir welding parameters on the microstructural and mechanical properties of AA 6016-T4 thin welds. Mater. Des. 1913–1921 (2009). http://dx.doi.org/10.1016/j.matdes.2008.09.016
18. C. Leitão, R.M. Leal, D.M. Rodrigues, A. Loureiro, P. Vilaça, Mechanical behaviour of similar and dissimilar AA5182-H111 and AA6016-T4 thin friction stir welds. Mater. Des. 101–108 (2009). http://dx.doi.org/10.1016/j.matdes.2008.04.045
19. E. Cerri, P. Leo, Warm and room temperature deformation of friction stir welded thin aluminium sheets. Mater. Des. **31**, 1392–1402 (2010). https://doi.org/10.1016/j.matdes.2009.08.044
20. K. Sithole, V.R. Veeredhi, Recent developments in micro friction stir welding—a review, in Accepted for presentation at the *2nd International Manufacturing Engineering Conference* (iMEC 2015), Kuala Lumpur, Malaysia (2015)
21. E.T. Akinlabi, R.M. Mahamood, S.A. Akinlabi, E. Ogunmuyiwa, Processing parameters influence on wear resistance behaviour of friction stir processed Al-TiC composites. Adv. Mater. Sci. Eng. (2014). http://www.hindawi.com/journals/amse/2014/724590/
22. A. Fehrenbacher, N.A. Duffie, N.J. Ferrier, F.E. Pfefferkorn, M.R. Zinn, Effects of tool-workpiece interface temperature on weld quality and quality improvements through temperature control in friction stir welding. Int. J. Adv. Manuf. Technol. **71**, 165–179 (2014)
23. X. Deng, X. Shaowen, Two-dimensional finite element simulation of material flow in the friction stir welding process. J. Manuf. Process. **6**, 125–133 (2004)
24. S. Tutunchilar, M.K. Besharati Givi, M. Haghpanahi, P. Asadi, Eutectic Al-Si piston alloy surface transformed to modified hypereutectic alloy via FSP. Mater. Sci. Eng. A **534**, 557–567 (2012)
25. A. Dolatkhah, P. Golbabaei, M.K. Besharati Givi, F. Molaiekiya, Investigating effects of process factors on microstructural and mechanical properties of Al5052/SiC metal matrix composite fabricated via friction stir processing. Mater. Des. **37**, 458–464 (2012)
26. M. Salehi, M. Saadatmand, J. Aghazadeh Mohandesi, Optimization of process factors for producing AA6061/SiC nanocomposites by friction stir processing. Trans. Nonfer. Met. Soc. China **22**, 1055–1063 (2012)
27. D.-H. Choi, Y.-I. Kim, D.-U. Kim, S.-B. Jung, Effect of SiC particles on microstructure and mechanical property of friction stir processed AA6061-T4. Trans. Nonfer. Met. Soc. China **22**, 614–618 (2012)
28. D. Aruri, K. Adepu, K. Adepu, K. Bazavada, Wear and mechanical properties of 6061-T6 aluminum alloy surface hybrid composites [(SiC + Gr) and (SiC + Al_2O_3)] fabricated by friction processing. J. Mater. Res. Technol. 362–369 (2013)

29. R. Dhayalan, K. Kalaiselvan, R. Sathiskumar, Characterization of AA6063/SiC-Gr surface composites produced by FSP technique. Procedia Eng. **97**, 625–631 (2014)
30. K. Schwab, *The Fourth Industrial Revolution* (Crown Publishing, New York, 2016)
31. V. Smil, *Creating the 20th Century: Technical Innovations of 1867–2014 and Their Lasting Impact* (Oxford University Press, New York, 2005)
32. T. Hughes, Technology as second creation, in *Human Built World* (University of Chicago Press, London, 2004), pp. 17–43
33. R. Chaâri, F. Ellouze, A. Koubâa, B. Qureshi, N. Pereira, H. Youssef, E. Tovar, Cyber-physical systems clouds: a survey. Comput. Netw. **108**, 260–278 (2016)
34. A.W. Colombo, S. Karnouskos, Towards the factory of the future: a service-oriented cross-layer infrastructure, in *ICT Shaping the World: A Scientific View*. European Telecommunications Standards Institute (ETSI) (Wiley, 2009), pp. 65–81
35. A.J.C. Trappey, C.V. Trappey, U.H. Govindarajan, J.J. Sun, A.C. Chuang, A review of technology standards and patent portfolios for enabling cyber-physical systems in advanced manufacturing. IEEE Access **4**, 7356–7382 (2016)
36. R.M. Mahamood, E.T. Akinlabi, Achieving mass customization through additive manufacturing, in *Advances in Ergonomics of Manufacturing: Managing the Enterprise of the Future*, ed. by C. Schlick, S. Trzcieliński (Springer International Publishing, Switzerland, 2016), pp. 385–390
37. S. Karnouskos, Realising next-generation web service-driven industrial systems. Int. J. Adv. Manuf. Technol. **60**(1–4), 409–419 (2011)
38. S. Karnouskos, A.W. Colombo, T. Bangemann, K. Manninen, R. Camp, M. Tilly, M. Sikora, F. Jammes, J. Delsing, J. Eliasson, P. Nappey, J. Hu, M. Graf, The IMC-AESOP architecture for cloud-based industrial cyber-physical systems, in *Industrial Cloud-Based Cyber-Physical Systems* (Springer, May 2014), pp. 49–88
39. A.W. Colombo, T. Bangemann, S. Karnouskos, J. Delsing, P. Stluka, R. Harrison, F. Jammes, J.L. Lastra (eds.), *Industrial Cloud-based Cyber-Physical Systems: The IMC-AESOP Approach* (Springer, 2014)
40. S. Karnouskos, V. Somlev, Performance assessment of integration in the cloud of things via web services, in *2013 IEEE International Conference on Industrial Technology (ICIT)* (Cape Town, South Africa, 2013). https://doi.org/10.1109/ICIT.2013.6505983
41. A.D. Maynard, Navigating the fourth industrial revolution. Nat. Nanotechnol. **10**, 1005–1006 (2015)
42. D. Mishra, R.B. Roy, S. Dutta, S.K. Pal, D. Chakravarty, A review on sensor based monitoring and control of friction stir welding process and a roadmap to Industry 4.0. J. Manuf. Process. **36**, 373–397 (2018)
43. M. Armbrust, I. Stoica, M. Zaharia, A. Fox, R. Griffith, A.D. Joseph et al., A view of cloud computing. Commun. ACM **53**, 50 (2010). https://doi.org/10.1145/1721654.1721672
44. B. Rossi, *Why Cloud Technology is Central to Industry 4.0*, pp. 6–8 (2018). http://www.information-age.com/cloud-technology-centralindustry-4-0-123462532/. Accessed 7 May 2018
45. F. Bonomi, R. Milito, J. Zhu, S. Addepalli, Fog computing and its role in the internet of things, in *Proceedings of the First Edition of the MCC Workshop on Mobile Cloud Computing* (MCC '12) (2012). https://doi.org/10.1145/2342509.2342513. 13
46. A. Parrott, W. Lane, *Industry 4.0 and the Digital Twin* (Deloitte University Press, 2017), pp. 1–17

Index

E. T. Akinlabi and R. M. Mahamood, *Solid-State Welding: Friction and Friction Stir Welding Processes*, Mechanical Engineering Series,
https://doi.org/10.1007/978-3-030-37015-2

MIX
Papier aus verantwortungsvollen Quellen
Paper from responsible sources
FSC® C105338

If you have any concerns about our products,
you can contact us on
ProductSafety@springernature.com

In case Publisher is established outside the EU,
the EU authorized representative is:
Springer Nature Customer Service Center GmbH
Europaplatz 3, 69115 Heidelberg, Germany

Printed by Libri Plureos GmbH
in Hamburg, Germany